Nature Strange and Beautiful

Nature Strange and Beautiful

How Living Beings Evolved and Made the Earth a Home

Egbert Giles Leigh, Jr.
Christian Ziegler

Yale
UNIVERSITY PRESS
New Haven & London

Published with assistance from the foundation established
in memory of Calvin Chapin of the Class of 1788, Yale College.

Yale University Press, in association with the Smithsonian Tropical Research Institute.

Yale University Press books may be purchased in quantity for educational, business, or promotional use. For information, please e-mail sales.press@yale.edu (U.S. office) or sales@yaleup.co.uk (U.K. office).

Set in Bulmer type by Tseng Information Systems, Inc.
Printed in the United States of America.

ISBN 978-0-300-24462-5 (hardcover : alk. paper)
Library of Congress Control Number: 2019931653
A catalogue record for this book is available from the British Library.

This paper meets the requirements of ANSI/NISO Z39.48-1992 (Permanence of Paper).

10 9 8 7 6 5 4 3 2 1

Contents

Preface

THE IDEA OF EVOLUTION BY NATURAL selection can heighten our appreciation of the beauty of nature. The French mystic Simone Weil believed that the true definition (and proper function) of science is the study of the beauty of the world; this book accordingly tries to show how evolutionary thinking can help us appreciate this beauty. We supply compelling evidence that the world's plants, animals, and microbes evolved and diversified from common microbial ancestors in response to natural selection: evolution by natural selection is indeed an appropriate lens for studying this beauty. To communicate this beauty, however, requires not only a scientist, but gifted artists and talented photographers, such as Christian Ziegler, the creator of so many of the stunning photographs in this book.

Thus we focus on the beauty and strangeness of nature, the unexpected steps and curious mechanisms by which the natural world we find so beautiful came to be. A steady energy subsidy through a crack in the sea bottom, in the floor separating the ocean from the volcanically turbulent underworld below, enabled beings that harnessed this energy to reproduce themselves, to arise from non-living processes through the coordination of a host of chemical reactions. Descendants of these multiplying beings transformed the geology near the earth's surface and took the first steps toward making the earth a suitable home for life. More than 500 million years ago, consumers of living beings appeared in an unprepared world, prompting an ever growing number of both predators and the algae- or detritus-eaters they preyed upon to evolve ever improving awareness of, coordinated with continually improving responsiveness to, events in their surroundings. This awareness was part of the foundation for the evolution of human minds. Indeed, habits evolved for one purpose sometimes have far-reaching consequences in startlingly different realms. Play presumably evolved in mammals, especially social mammals, as exercise in activities requiring close coordination between eye, mouth, and foot, and as practice among playfellows in social relations. In some animals, play became a form of disinterested exploration of their worlds, learning for learning's sake. Science and art are forms of play that, in the right hands, attained the immaterial truths of mathematics and the glorious achievements of Dante's *Commedia,* Einstein's general theory of relativity, and Paul Dirac's quantum mechanics. A sense of beauty drove all these achievements, which in turn enhanced our appreciation of the beauty of the world.

Natural selection is competition. Yet in the natural world, as in human economies, organisms cooperate to compete better with third parties, as the economist Adam Smith was the first to show. Thus a tropical forest, like a modern human economy, is both an arena of intense competition and an acme of interdependence and cooperation. As Smith also argued, rules of fairness must be enforced if competition is to serve the common good. Thus Plato's dictum—that a gang of thieves is effective only if its members treat each other justly—helps us see why an animal's (notoriously selfish!) genes collectively enforce rules ensuring that a gene spreads only if it benefits the individuals that carry it. More generally, the ways animals in a cooperative group prevent cheating that undermines all the good of cooperating sheds light on why cancerous cell lines so rarely spread and kill the animals they belong to. Indeed, the analogy between maintaining cooperation within the company of an animal's cells and maintaining cooperation in the company of bees in a honeybee hive has shed light on how activities of an animal's cells, and a hive's honeybees, are coordinated. Similarly, the Dutch philosopher Spinoza's dictum that our own self-interest requires that we benefit those who benefit us sheds light on how the great grazers of the Serengeti grasslands protect these grasslands by keeping trees out and enriching the grassland by manuring the grass appropriately, thus making these grasslands a better home for grazers.

Abundant experience of nature at its most beautiful—especially in the tropical forest of Barro Colorado Island in Panama, but also in Madagascar, Peru, and other tropical countries, as well as the rocky wave-beaten shores of Tatoosh, a cluster of islets off the northwest tip of the Olympic Peninsula in Washington—is reflected in this book. It has been written and rewritten in an office on Barro Colorado, inspired by the music of Bach, Monteverdi, and many others. This book was written in collaboration with Christian Ziegler, who likewise is thoroughly familiar with Barro Colorado Island and nearby mainland in Panama, but who also has worked in the Democratic Republic of the Congo, northeast Australia, Borneo, Thailand, Bhutan, and many other exotic sites. This book is a counterpart about evolution to *A Magic Web,* a book about the ecology and natural history of Barro Colorado that Christian and I collaborated on some years ago. By working together once again, we hope this book will help to show how understanding evolution reveals the beauty of nature.

Acknowledgments

FIRST AND FOREMOST, WE ARE MOST grateful to the artists and other photographers who contributed to this book. The contributing artists are Debby Cotter Kaspari, Damond Kyllo, Anne Klein, Barrett Klein, Mary Bruce Leigh, Aaron O'Dea, and Danielle VanBrabant; the contributing photographers, Anne Paine, Robert Delfs, Janie Wulff, Annette Aiello, Patrick Kennedy, and Teague O'Mara. Next, we heartily thank those who read successive drafts of the whole manuscript: Anthony Coates, Henry Horn, and Richard Schooley. Janie Wulff provided helpful comments on parts of the book. Various schoolteachers visiting Barro Colorado from the Professional Resources Institute for Science and Mathematics, Montclair State University, New Jersey, read the chapter on heredity and natural selection about elementary genetics, and declared it readable (when it was still devoid of illustration!). Christie Henry, formerly with the University of Chicago Press, provided remarkably helpful and constructive advice while the work was still in progress. Karen Kostyal greatly improved the beginning of the book, especially the section on the origin of life. Finally, I am deeply indebted to Geerat Vermeij, with whom I have discussed the problems of evolutionary biology so often that it is difficult to discern whose idea was whose. His astounding command of the literature has plugged many a gaping hole in my knowledge.

We are both most grateful to the various communities at the Smithsonian Tropical Research Institute. The intellectual community provided and continually refined our education in tropical biology, from the contrasts between the two very different oceans flanking the Isthmus of Panama, the significance of which Charles Birkeland, Janie Wulff, and Peter Glynn did so much to reveal, to the many different aspects of tropical forest, whose organizing principles were first elicited by a graduate student, Robin Foster. Moreover, work in Panama has set this knowledge in the context of a history extending from the age of the dinosaurs to the present. Martin Moynihan, the institute's founding director, gave me my start in tropical biology; his successors, Ira Rubinoff and Eldredge Bermingham, helped Christian Ziegler's career by financing the publication of successive editions of *A Magic Web,* a photographic essay on the forest community of Barro Colorado Island, and the current director Matthew Larsen provided financial support for publishing this book. We also thank the support staff of the Smithsonian Tropical Research Institute, particularly Adriana Bilgray, whose management of fellowship programs does so much to maintain the quality of the institute's intellectual life, and Oris Sanjur, Oris Acevedo, and Melissa Cano, who have done so much to keep science going on Barro Colorado Island by making it so agreeable and easy to work there.

Finally, we thank Yale University Press, especially Jean Thomson Black, the science editor, Michael Deneen, editorial assistant, and Phillip King, the book's copy editor, and also other branches of the Press that authors only hear of second-hand, for the concern and loving care they have lavished on this book.

EGBERT GILES LEIGH, JR.
CHRISTIAN ZIEGLER
Feast of St. Matthew, tax collector,
apostle and evangelist, 2018

Nature Strange and Beautiful

ONE

Introduction

NEARLY FOUR BILLION YEARS AGO, the first living things—systems that reproduced themselves with energy harnessed from their surroundings—began to multiply in obscure places on the sea bottom, perhaps a kilometer deep. They apparently arose from lifeless matter. (If this were not so, science could not reveal how life began!) No living being, however, is simple. Even in the simplest beings, using chemical energy to grow and multiply requires controlling and coordinating the place and timing of many hundreds, if not thousands, of chemical reactions. To do this, such beings make catalysts—enzymes—that specify what reactions occur, and they make other chemical compounds that regulate when and where these enzymes are made and coordinate the activities of different enzymes. Specifying, controlling, and coordinating these chemical reactions laid the foundation for the "natural technology" that living beings use to eat, grow, multiply, and cooperate. This achievement enabled life's origin.

Duplicating an organism's *genome*—its instructions for making its enzymes and coordinating their activities—inevitably entails copy errors that create differences among its descendants, some of which affect their ability to multiply. Such differential reproduction, which we call natural selection, causes those variants that multiply fastest to spread at their fellows' expense. An organism's genome may be viewed as its hypothesis, encoded in its DNA, of how to live and reproduce in its environment. By trial and error, strictly speaking, by testing variant hypotheses generated by copy errors, natural selection improves this hypothesis, coordinating the organism's function more nearly with relevant features of its environment. Similarly, a person improves a scientific hypothesis, conforming it more nearly to the truth, by testing successive modifications through experiment or observation. Natural selection, however, is a mindless mechanism, the automatic process of differential reproduction. Moreover, the environment, unlike truth, changes.

Once reproductive entities appeared, better reproducers replaced worse. This natural selection transformed their descendants into living beings, ever more clearly organized for making a living to reproduce their kind. Being able to multiply set these beings—these organisms—and their planet on the road to radical change. By modifying their surroundings, making them more hospitable to life, they transformed the planet as well. Recent descendants have evolved that are capable of conceptual thought.

In this book, we ask where and how self-reproducing beings might first have appeared, how they created so great a diversity of organisms and landscapes, and how some evolved greater awareness of, and ability to react to, their surroundings. What innovations allowed large trees and active animals to evolve? How can tall trees raise water and nutrients to their topmost leaves, conserve water in time of drought, "know" when to drop old leaves, and when to flower or flush new leaves? How can different animals see, hear, smell, or feel food and predators, move accordingly to eat appropriate food and avoid predators, and use their food to fuel their activities (plate 1.1)? Cats and monkeys, for instance, possess unconscious computational skills that enable their eyes and brain to abstract an object from its various perspective views, and infer whether it is a predator to avoid, an obstacle to circumvent, or prey to catch and eat. As cells of a many-celled animal send each other signals to coordinate their activities, so social animals communicate with each other to coordinate their group's activities.

As living things spread and diversified, they associated in ever more diverse communities where, to live and reproduce, each member depended on the activities of others of many different kinds, just as a modern city-dweller depends on others of many different occupations in order to live and raise a family. How did living things become so diverse? Why did natural selection favor the relationships of interdependence and cooperation that led to the luxuriance and diversity of coral reefs, rain forests, and the great African grasslands? Indeed, species with very different abilities often pool them for their mutual advantage, like the insects that carry pollen from one plant to another to obtain the special nectar that the immobile plant provides in return. As in human economies, the luxuriance and diversity of natural ecosystems depends on cooperative endeavor. How can cooperators find and help each other, without being cheated?

An ultimate animal technology is conscious human minds, endowed with the capacity for conceptual thought, moral judgment, and language. Human minds enabled cultural change—the spread of ideas and practices learned from other human beings—to replace natural selection as the primary driver of change in human social behavior and the ways human beings make their livings. Human minds spawned the technology allowing the intricate social cooperation that transformed, and often marred, the earth with unparalleled speed. How could consciousness, conceptual thought, and language evolve?

This book is not a comprehensive review of evolution and how it happened. The subject is too vast, and reviewing each topic in equal detail would be too dull. Like that philosophical physicist Hermann Weyl, whose books have been an inspiration since undergraduate days, we "prefer the open landscape under a clear sky, with its depth of perspective, where the wealth of sharply defined nearby details gradually fades away toward the horizon." This book is a personal view, based on fifty years of thinking about how natural selection works, fifty years of experience in tropical forests from Panama to Madagascar and Malaysia, and talks with many a student of forest, coral reef, and rocky shore. It focuses on the extraordinary coordination among an organism's different parts and processes and between organism and environment, how this coordination came to be, and how the competitive process of natural selection can lead to social cooperation within species and mutualism among species. Both author and photographer favor examples we know personally: hence the emphasis on Panama, where we have long worked at the Smithsonian Tropical Research Institute, surrounded by colleagues and students who helped to educate us. Any book on evolution, however, must seek examples far beyond one small island or one country.

TWO

How We Approach the Problem

LIFE INVOLVES COORDINATING different processes and activities, at many different levels. How did these different sorts of coordination arise? The ability to coordinate numerous activities at many levels allows a living being to procure the means to live and reproduce in the habitat to which it is adapted (plate 2.1). We marvel at how a cat's form, physiology, and behavior are suited to catching prey, or a camel's to living on desert plants. We seldom wonder how intricately their metabolic processes, and activities of nerve and muscle, must be coordinated for animals to procure energy and deploy it to survive and multiply. Even a bacterium must specify and coordinate thousands of chemical reactions to harness the energy needed to live and multiply. Likewise, a many-celled animal must coordinate the activities of its multitude of cells, and a society of many-celled animals, the activities of its members. Ecological communities, like human ones, are webs of interdependence, not just arenas where individuals compete for food.

To flesh out what we wish to explain, consider a female ocelot that infers a multi-modal "movie" of selected features and events in her surroundings—a real-time hypothesis of the environment she lives in—by coordinating impressions from sight, smell, touch, and hearing (plate 2.2). Using this picture of what is happening around her, the ocelot organizes appropriate actions by which she can detect, catch, and eat prey, avoid being eaten, find safe places to sleep, choose suitable mates, and bear and raise her young. The food she eats must be digested, and distributed, along with oxygen, among the ocelot's many cells, and wastes removed and expelled from the animal. On the other hand, a cyanobacterium "knows" far

Fig. 2.1. An individual army ant is marvelously stupid, but a colony of 500,000 behaves adaptively. On Barro Colorado Island in central Panama, these army ant workers, *Eciton burchelli,* are making their colony's nest of their own bodies. They will keep the temperature inside constant within 1° C. (Photograph by Christian Ziegler)

less about its surroundings, and has a different, far more limited set of possible responses, but it uses the power of sunlight to drive a set of complicated, intricately coordinated chemical reactions that turn carbon dioxide and water into carbohydrates providing the power it needs to survive and multiply. Using these carbohydrates involves many chemical reactions: each must occur in the right places, at the right times. Extracting and deploying energy from the food an ocelot eats to power its growth and activities involves coordinating even more reactions, even more intricately. A honeybee society has one reproductive queen with several tens of thousands of daughters who, instead of reproducing on their own, spend their lives helping her reproduce by nursing her young, maintaining the nest, looking for and gathering suitable food, and telling other foragers where to find it. Each honeybee worker must detect and infer, from sight, smell, and hearing, relevant features and events in her surroundings, and coordinate the responses needed to fulfill her tasks. But, just as an ocelot must regulate her eating to match her needs, so must the society of honeybees. More generally, all the activities of its many workers must be timed and coordinated by appropriate signals so as to enhance the queen's reproduction.

Indeed, there are many levels of coordination in nature—among a cell's chemical processes, among the activities of a many-celled animal's different cells, tissues, and organs, and among those of the different individuals in a social group (fig. 2.1). Each level's coordination represents adaptation—the organization of a bacterium, a many-celled plant or animal, or a social group, to survive and reproduce. Each level's coordination depends on prior adaptation in its component parts and processes: higher levels depend on functional lower levels. Moreover, just as our human society depends on coordination of many activities ranging from producing and distributing food to constructing and maintaining the communications networks by which the society's other activities are coordinated, so a natural community depends on coordination of activities of its different kinds of plants, animals and microbes.

This book starts by demonstrating adaptation—in visible form, color, and behavior of selected plants, animals, and animal societies, and pointing out the intricacy of some of the various types of coordination that allow living things to function. How are their structures and activities organized and coordinated to procure sufficient resources from their habitats to live and multiply? Living beings are adapted to particular habitats: adaptation coordinates an organism's structure, processes, and behavioral repertoire with its habitat. Moreover, group life enables some animals to make livings in ways no lone animal could do. How do a social group's members coordinate their activities?

Then we ask why, and in what ways, individual and social adaptation evolved. First, we briefly summarize evidence that all living things have diversified from shared common ancestors. Then we suggest how life began—how beings that harnessed energy from their surroundings to reproduce themselves arose from lifeless matter. This achievement brings us face to face with how many chemical processes must be controlled and coordinated to allow the simplest beings to grow and multiply. Once this happened, those that reproduced better supplanted the others. This natural selection conferred on these beings a purpose in life, self-reproduction, that made them so different from lifeless matter that it was long believed that living things, with their purposeful, responsive behavior, could never have arisen from matter.

Next, we ask why living beings have diversified, forming communities, each with interdependence, and often crucial cooperation, among various of its member species. In human societies, new technology provides more, and more specialized, ways to make livings, allows the coordination of more complex forms of cooperation on ever wider scales, and enhances economic productivity. Indeed, just as a modern human city-dweller depends on others with many different jobs to procure the means to live and raise a family, so any plant or animal depends on other organisms of many different kinds in order to live and reproduce.

Life in a community requires many natural technologies. How were they improved and diversified? Early microbes catalyzed reactions between different chemicals to procure the energy and make the building blocks they needed to grow and multiply, leaving other chemicals as by-products, some of which provided livings for other microbes. As chemical technology improved and diversified, primitive photosynthesizers evolved that used light to drive sugar-producing chemical reactions. Some of them left sulphur deposits, others, massive deposits of iron ore. Finally, by 2.7 billion years ago, cyanobacteria evolved "oxygenic" photosynthesis, a marvel of intricately coordinated technology that transformed water and carbon dioxide into sugars and oxygen. These bacteria oxygenated the earth's atmosphere and began to oxygenate the surface waters of the oceans. These microbial populations, and many others, whose members were naturally selected to multiply rapidly, transformed the earth into a better home for an enormous variety of living beings. Photosynthetic oxygen—a poison where it first appeared—eventually allowed the evolution of active animals and large plants, many of which could coordinate different activities with each other in mutually profitable ways like plants and their pollinators or honeybees in a colony.

Almost two billion years ago, a microbe engulfed others of a very different kind, some of which managed to survive and reproduce within their hosts. These live-in bacteria posed their hosts problems that were nearly fatal. The hosts succeeded in improving their own prospects by co-opting their guests' metabolic abilities. To do so, however, the hosts had to evolve orderly sexual reproduction, and change in ways that made diversification much easier, transfer of genes among species (for bacteria, a major source of new variation) much rarer, and mutual adjustment of an interbreeding population's genes much closer. Their descendants were "eukaryotic" cells, with possibilities none of their predecessors shared. How did selection transform this relationship into the most successful partnership in all the history of life? This event opened the way for a great variety of one-celled organisms, and multicellular animals, fungi, and plants to evolve. Clonal groups of these microbes evolved systems of signaling between cells that enabled division of labor among them, leading to the diverse array of complex animals and plants that populated the earth during the last half-billion years and more. How are the multiplication of cells in these clumps, and their various activities, coordinated and controlled to produce functional individuals? How did this coordination evolve? Such technologies allowed microbes, plants, and animals to spread to new places, make their livings in new ways, and develop new ways to cooperate.

Having discussed distinctive characteristics of life and some major events in its evolutionary history, we next consider how all this could arise. First we discuss genetics—how parents pass on characteristics to their offspring. The DNA of an organism's genes encodes the processes and abilities that enable it to grow, develop, and reproduce. Collectively, these genes, the organism's genome, encode the organism's hypothesis of how to make a living in its environment. How are genetic systems tailored to allow natural selection to keep rogue genes from multiplying at the expense of their bearers' welfare, and to favor adaptive evolution most effectively—to allow a population's hypothesis of how to make a living to improve most rapidly? Now, genes of multicellular organisms are transmitted by rigid rules ensuring that new mutants spread only if they benefit their bearers. Moreover, sexual reproduction enhances the likelihood that a new mutant spreads according to its own merits, rather than by the quality of the genome in which it first arose. How did natural selection among unconscious organisms such as plants and flies yield and enforce such rules of fair competition—rules that ensure that competition between genes benefits their genome (complex modern societies still struggle to achieve an organization whereby competition among individuals benefits society)? Even with rules ensuring fair competition between genes, how could natural selection bring forth individual and social adaptation, diversification, and cooperation among different species? Cooperation offers opportunities for cheating, benefiting from the work of others while contributing nothing in return. Cheating can make cooperation, either within or among species, a waste of effort. How can natural selection limit cheating? Finally, how could adaptation, diversification and cooperation make the earth's habitats more suitable for life and generate ecosystems—complex natural economies? Complex, productive ecosystems require new technologies enabling more effective individual and social adaptation, and enhancing ecosystem function. One such technology is the ultimate tool for coordinating both an individual's activities and cooperation among individuals—conceptual thought and language.

THREE

Adaptation, Individual and Social

LIFE'S MOST CHARACTERISTIC FEATURE is that the forms, colors, functions, and behaviors of living beings are mutually coordinated—adapted—to let them acquire the resources they need to live, grow, and multiply. A 50-gram (2 ounce) female fruit bat can fly, use "sonar" to navigate in the dark, use sonar and smell to find ripe fruit she can eat and digest, remember locations of fruiting trees, choose mates, and produce and care for young. These abilities enable the bat to find and eat ten or twelve figs of 70 grams total weight in a night, from which she derives a power supply of 0.8 watts, enough to allow her to live, move, and reproduce (fig. 3.1). Human beings cannot make anything with this array of abilities, let alone a robot that can do all this on a mere 0.8 watts.

Darwin realized that a plant or animal whose anatomy, color, and behavior enabled it to survive and reproduce better than its fellows would leave more offspring in the next generation. If a population's members differ in ability to reproduce, natural selection happens: the most reproductive leave the most offspring. Thus, barring adverse environmental change, each generation is as good as or better than its parents at making a living from its habitat. The coordination among an animal's characteristics that allows it to practice its way of life in some habitat well enough to survive and reproduce is called adaptation. Adaptation is therefore a relation between the forms and behaviors of living things, their ways of life, and their environments.

Individual Adaptation

How do we recognize adaptation? In 1930, Ronald Fisher, a founder of mathematical statistics who wrote a fundamental book on evolution by

Fig. 3.1. This Panamanian bat, the 50-gram *Artibeus jamaicensis,* lives and reproduces on a power supply of less than one watt, which it derives by eating figs like those borne by the fig tree, *Ficus insipida,* shown here. (Photograph by Christian Ziegler)

natural selection, defined adaptation as a relationship between an organism's characteristics and its environment: "An organism is regarded as adapted to a particular situation, or to the totality of situations which constitute its environment, only in so far as we can imagine an assemblage of situations, or environments, to which on the average the animal would on the whole be less well adapted; and equally only in so far as we can imagine an assemblage of slightly different organic forms, which would be less well adapted to that environment." This definition develops an argument of Aristotle that, since "sports," mutant individuals that differ from the normal in some way, usually survive and reproduce less well than normal counterparts, organisms are designed to survive and reproduce.

Animals must avoid being eaten. Edible insects do so by mimicking inedible objects. This mimicry is precisely adapted. The elaborate ways many insects, especially tropical ones, mimic things they are not, suggest what intense predation they face. Too imprecise mimicry can be fatal. Many insects deceive would-be predators by resembling sticks or leaves, which these predators do not eat (plate 3.1). Experiments in Panama showed that small monkeys, visual predators, detect insect prey by their heads and paired legs. Therefore, resting walking sticks (phasmids) extend their juxtaposed forelegs to simulate the forepart of a stick, covering their heads, eyes, and antennae: the forelegs have a special notch in which the head fits. Their other legs are kept under their bodies, or folded to look like short stubby shoots from a stick.

Other animals, even some frogs, avoid being eaten by advertising their distastefulness with bright colors. Distasteful species of butterfly, such as passionvine butterflies, *Heliconius,* may mimic each other so that local predators need learn just one color pattern to avoid. This mimicry represents cooperation between members of different species to reduce the cost of teaching potential predators that they taste awful. In Panama, *Heliconius melpomene* mimics the equally distasteful *H. erato.* In both, wings are black, with red crossbars across their forewings and yellow leading edges on their hindwings. They both frequent roadsides and brightly lit clearings, looking a bit like little black boomerangs with a red crossbar near each tip, joined by a yellow stripe along the boomerang's center. *Heliconius cydno* resembles *H. sapho;* they both have blue-black wings with white crossbars on their forewings, and live in shady forest interiors. *H. cydno* diverged only recently from *H. melpomene,* whereas *H. erato* is more closely related to *H. sapho* than to the two others. Nonetheless, the manner of flight, like the wing colors, of *H. erato* are far more like *H. melpomene*'s than its closer relative's, whereas *H. cydno*'s manner of flight, and its wing colors, are most like *H. sapho*'s. Butterflies with the same color pattern share the same flight behavior. This resemblance matters: hybrids with unusual colors and behavior die twice as fast as butterflies with colors and manners of flight that predators recognize.

Many edible insects cheat, however, by mimicking distasteful ones. The nastier the model, the more precise the resemblance. The rarer the mimic relative to the model (1/20 or less of the model's abundance, as a rule), the less likely predators are to notice and eat the edible mimic. In Panama, black butterflies with green or cream-colored spots on their forewings and red ones on their hindwings visit flowers in sunlit gaps and along forest edges. Most are species of the genus *Parides:* their caterpillars resemble feces, eat the poisonous leaves and flowers of pipevine, *Aristolochia* spp., and retain the poisons in their own bodies into adulthood. A few of these black butterflies with bright spots are edible *Mimoides ilus,* which pretend to be poisonous by mimicking poisonous *Parides* (plate 3.2). Indeed, they cheat the *Parides* by undermining the reliability of the warning conveyed by *Parides*'s colors, but they are too rare to annihilate the warning colors' usefulness. Mimicking the same inedible object too often can make it easier for predators to recognize the disguise. Thus the common leaf-backed toad of Barro Colorado's forest floor, *Rhinella alata,* mimics dead leaves in many different ways, lest an oft-repeated disguise attract attention (plate 3.3).

Plants must also avoid being eaten. The anti-herbivore poison content of leaves of the understory shrub *Psychotria horizontalis* is adjusted to achieve the best balance between excess leaf damage and overspending on leaf defense. They defend their leaves against herbivores with tannins like those used to preserve (tan) leather. Leaves with more tannin suffer less herbivore damage, but cost more to make. The happy medium that promotes maximum net growth is that tannin level where the reduced growth from making a little more tannin just balances the decrease in herbivory that results. Cynthia Sagers, a graduate student on Barro Colorado, asked whether these *Psychotria* strike this optimum level in tannin content. Since *Psychotria* cuttings take root if stuck in the ground, she took two cuttings each from twenty different *Psychotria horizontalis.* She rooted one set in the forest understory, protected from insects by a mesh cage, and rooted the others nearby, but exposed to insects. She harvested these cuttings after a year. Exposed cuttings showed no correlation between a cutting's weight gain over the year and its leaves' tannin content. Those plants whose leaves had more tannin were less eaten, but this advantage just balanced the slower growth resulting from making more tannin—what one would expect were their tannin content the most appropriate for their habitat. Protected cuttings whose leaves had the most tannin gained the least weight, because they had too much tannin for the low herbivory in their protective cages.

Large active animals must bring oxygen and food to their tissues, and remove their carbon dioxide and other wastes. This problem is most acute for warm-blooded mammals and birds: their cells need lots of oxygen. An 80-kilogram (175-pound) man at rest consumes energy at the rate of 80 watts, compared with 20 for a resting cold-blooded 80-kilogram alligator. To feed its cells, a bird or mammal's heart pumps blood through its arteries and capillaries. Once the blood has delivered its food and oxygen to the animal's cells and become loaded with carbon dioxide and other wastes, it returns through the veins to the heart, discarding wastes and becoming recharged with food en route. The blood is then pumped through the lungs to discard carbon dioxide and take on oxygen. Then it flows back to the heart, which pumps it out anew to feed the tissues. An 80-kilogram man contains 8,000 square meters (two acres!) of capillary wall, and 100,000 kilometers (60,000 miles, more than enough to circle the world twice) of blood vessels, mostly capillaries—1¼ kilometers (¾ mile) of capillaries for every cubic centimeter, or 1¼ meters (4 feet) for every cubic millimeter, of flesh (fig. 3.2). The less power needed to circulate blood, the more power available for other activities. Are our circulatory systems organized to minimize the power needed to circulate blood from the heart to the capillaries and back? If so, how?

This problem looks fearsome. Nonetheless, in 1926 Cecil Murray, a professor at Bryn Mawr College, cracked it by tackling a simple but crucial piece of it. He asked what radius of blood vessel would minimize the power (energy per unit time) needed to move Q cubic centimeters of blood per second through it, plus the power needed to maintain the vessel and the blood in it. Narrowing the vessel disproportionately increases the power needed to move this much blood through it. Too big a vessel makes maintaining the blood in it too costly, even though this blood moves through it easily. Between these extremes is an optimal vessel radius, a happy medium, where the reduced cost of moving the blood caused by a slight increase in vessel radius just balances the increased cost of maintaining the extra blood in the enlarged vessel. At this optimum, the average speed of the blood moving through a vessel is proportional to its width, and the average volume of blood moving through it per second is proportional to the cube of its width: this is now called Murray's Law. In mammals like ourselves, this

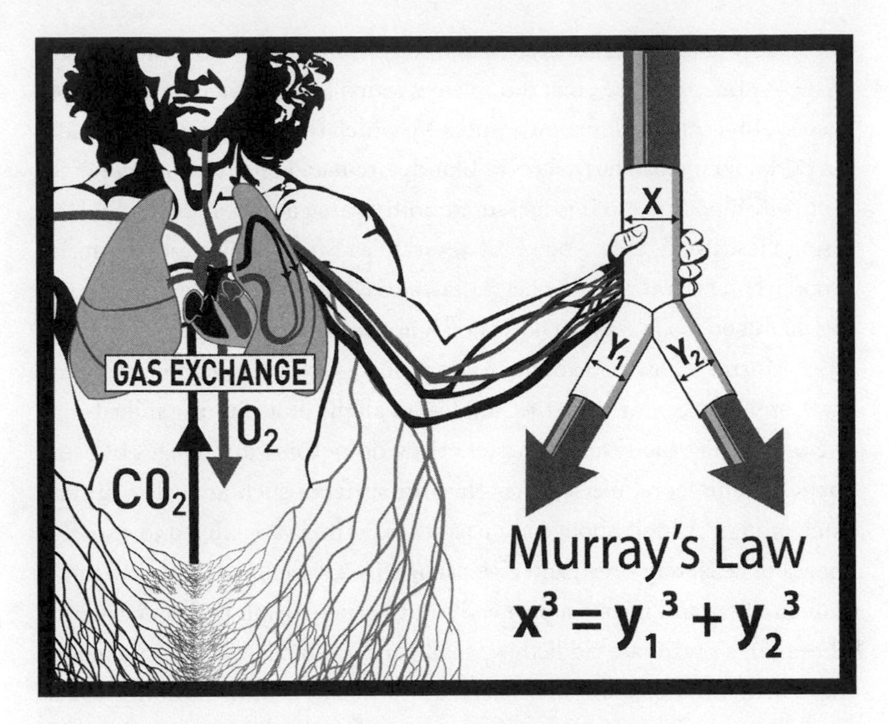

Fig. 3.2. Human beings have trillions of cells. All must be supplied with food and oxygen, and have their wastes removed. To do this, the heart pumps five liters of blood per minute through an adult's lungs to exchange carbon dioxide for needed oxygen, and then through a network of arteries to the rest of body, and from thence back to the heart through the veins. The combined cost of driving the blood through this system, and maintaining and renewing the blood it contains, is minimized by making the cubes of the diameters of the two branches equal the cube of the diameter of their parent where a vessel forks, as in the forked vessel in the man's left hand. The velocity of blood in a vessel is therefore proportional to its diameter. (Diagram by Damond Kyllo)

is nearly true for arteries of all sizes, but not for capillaries, which are so small that a blood corpuscle is as wide as the capillary itself, so blood does not flow through capillaries like a normal fluid. In mammals, therefore, the circulatory system is organized to circulate blood nearly as cheaply as possible. Moreover, the friction of blood against the vessel wall is the same for vessels of all widths. As cells in the vessel wall probably sense this friction, growth of arteries may be organized to keep the friction of blood against their walls constant.

How much power is needed to circulate a person's blood through his tissues? Murray realized that the power needed is the volume circulated per second, times the pressure differential by which the heart moves the blood. An 80-kilogram man has 5 liters of blood, circulated once a minute when he is at rest. Blood pressure is measured on the large artery in the upper arm. Using these two figures, a power of 4/3 watt is needed to circulate this man's blood when he is at rest. The heart draws 10 watts for every watt spent moving the blood, so it uses 1/6 of this resting man's 80-watt power demand.

Murray's Law also applies to transport systems of very different animals: sponges. Sponges are the simplest of all the multicellular animals that live by filtering food from seawater. They do not move. Sponges of many shapes live on coral reefs. Others live on surfaces such as rock walls and dock pilings. A basic sponge is a narrow-mouthed vase affixed to a rock—a mass of cells, reinforced and given shape by a finely branched skeleton of protein fibers and microscopic needles or branched spines called spicules. The sponge's walls are riddled by small canals and chambers. Water enters the sponge through small holes in its surface, and is drawn through canals branching into ever smaller channels to "flagellar chambers" 0.03 millimeter wide, lined by "collar cells," choanocytes, whose whiplike flagella move this water and snare microbes from it. A typical sponge has 10,000 such chambers per cubic millimeter of sponge wall. Filtered water is moved through channels that join into larger channels into the sponge's central chamber or atrium, from which it is expelled through the sponge's osculum (fig. 3.3). A larger sponge is more complex, but it still lives by moving water through canals, filtering it in flagellar chambers, and expelling it through

Fig. 3.3. (*opposite*) A sponge with a sector cut out, exposing its system of canals, and a detail of the exposed surface showing the canals and the flagellar chambers from which cells extract food from the moving water (A); a flagellar chamber (B); and a section of chamber wall showing the collar cells whose whiplike flagella, evolved originally for locomotion, beat the water, driving it through the sponge (C). Encircling the flagellum is a collar of microvilli that snare food particles from the water. A sponge's canals, like a vertebrate's arteries and veins, obey Murray's Law, thereby minimizing the combined cost of pumping this water and maintaining the canals through which it is pumped. (Drawing by Damond Kyllo)

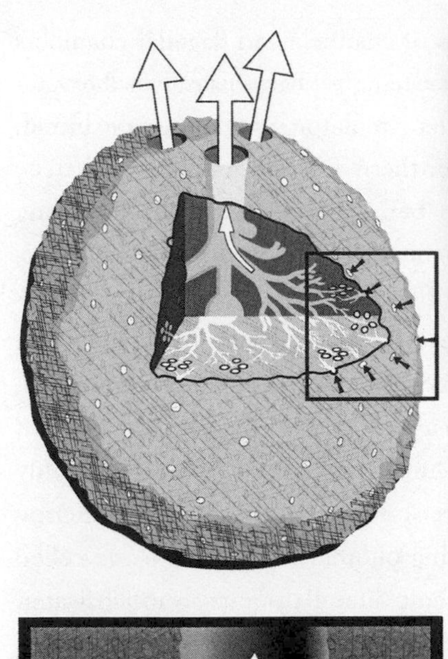

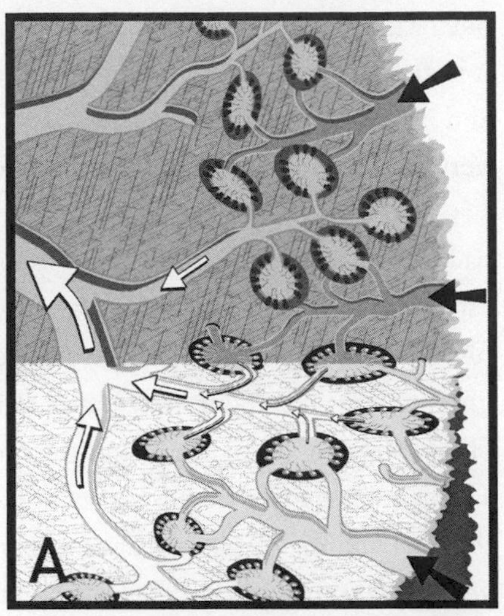
A

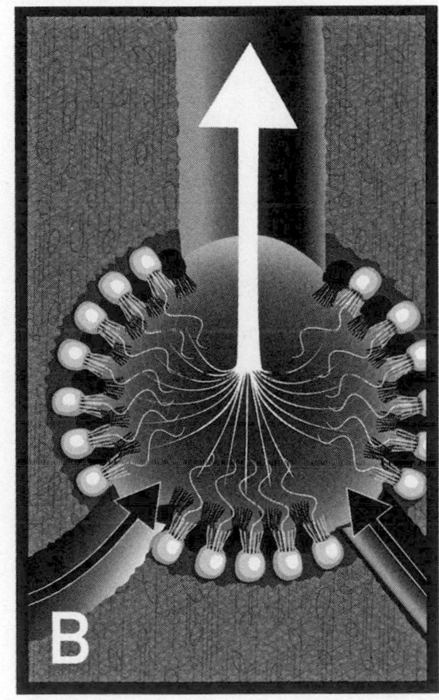
B

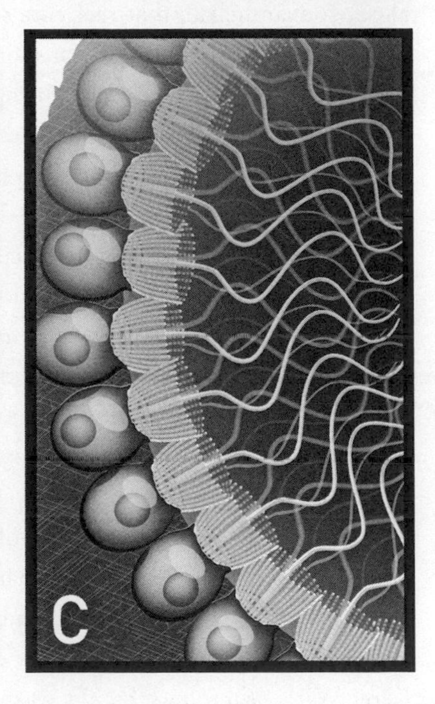
C

one or more oscula. A sponge's system of channels and flagellar chambers is its feeding apparatus, its respiratory system, its heart that moves the water and its food through all its parts, and its circulatory and excretory system. Vertebrates use very different organs for these different functions.

Sponges avoid being eaten, and being infected by disease-causing bacteria, by making extraordinary poisons that have become of great interest to drug companies. These poisons often let filter-feeding sponges multiply, little troubled by predators or disease, until there are only just enough microbes in the water to maintain their numbers. A graduate student, Henry Reiswig, found that sponges on a reef foreslope near Discovery Bay, Jamaica, filter three times their own volume of water every minute, roughly the total volume of water over their part of the reef every day. Their filtering keeps reef waters crystal clear, enhancing photosynthesis in the one-celled algae, zooxanthellae, that reef corals cultivate. To pump so much water, sponges need efficient water-transport systems. Water is free, but the cost of maintaining a canal's walls is proportional to the volume of water the canal contains. Therefore, a sponge's water channels obey Murray's Law as closely as do a person's blood vessels. The average velocity of water in a canal is proportional to its width, and the volume of water moving through it per second is proportional to the cube of its width.

Adaptation sometimes demands flexible behavior. In Panama, red-eyed tree frogs, *Agalychnis callidryas,* hatch as tadpoles in ponds with abundant plankton to eat and an abundance of shrimp and fish that eat tadpoles, later becoming terrestrial adults (plate 3.4). Adult female frogs come down from the tree crowns where they normally live to lay a gelatinous mass of 40 eggs on a leaf hanging over a pond. Hatching tadpoles drop into the water below. Their eggs face a trade-off. Later-hatching eggs produce larger tadpoles that escape the pond's predators more easily, but snakes and wasps have more time to find and eat them before they hatch. Normally, tadpoles hatch seven to nine days after their eggs are laid. As a graduate student, Karen Warkentin found that if a wasp or snake attacks eggs in a clutch five or six days after they are laid, tadpoles will hatch right away from uneaten eggs (plate 3.5). The immediate risk of being eaten outweighs greater long-term vulnerability to predators in the pond. She also showed that embryonic tadpoles know when a predator is attacking by how their leaf vibrates during an attack, and drop into the water. Vibrations from heavy rains do not prompt premature hatching. In sum, adaptation is reflected in many features of form, color, physiology, and behavior.

Social Adaptation: Why and How Do Group Members Cooperate?

Many animals live in social groups (plate 3.6). Group life offers many advantages. Some live in groups to cope better with predators or competitors: a group has more eyes that might notice predators, and more members to fight off predators or competitors too strong for one alone. Tropical forest birds of several species flock together when looking for food, because a flock has many pairs of eyes to watch for predators. Other animals, such as monkeys, peccaries, female coatis, many bats, wasps and bees, and all ants and termites live in groups all their lives. Bats that roost and forage together show each other what is edible and where to find food. White-faced monkeys live in groups: each group tries to keep others out of a territory big enough to assure its monkeys enough food all year long.

These animals all live in groups either because group life enhances their own reproduction, or because the benefits of helping one's mother (or other relative) reproduce exceed those of producing young of their own. Such behavior spreads those genes inherited from one's mother that favor helping her. Indeed, honeybee workers, which normally have no offspring, spread their genes only by helping the queen, their mother, reproduce. This circumstance suggests that adaptation serves the common interest of an organism's genes, rather than the good of the individual carrying them.

Social life poses challenges. How can group members tell each other of opportunities and dangers, and coordinate their activities? How do they keep their colonies healthy? How do they keep from being cheated by fellow group members?

HOW DO HONEYBEE COLONIES SURMOUNT THE PROBLEMS OF SOCIAL LIFE?

Honeybees are major pollinators in the Old World Tropics and the Americas. When summer begins, a nest of honeybees in upstate New York has roughly 30,000 bees, weighing 4 kilograms in all. Cells must be cleaned, the queen and brood cared for, the nest maintained, and nectar and pollen gathered into the nest. The queen influences her colony's division of labor only by volatile chemicals that signal her presence and good health. Honeybees' "technology of social life" governs distribution of tasks among workers, finding food, communicating its location to nestmates, and the like. How can honeybees do this without a central coordinator?

Partitioning tasks among workers is a major challenge in honeybee life. After emerging from its brood cell, an adult bee progresses through a series of tasks as it ages, beginning with tasks in the nest, which are the least risky. Newly emerged bees clean brood cells. Two days later, they become nurses, caring for the queen and her brood. Normally, they become food storers when 12 or 13 days old, accepting nectar from returning foragers and storing it, building new comb when it is needed to store incoming food, etc. They become foragers, the riskiest task of all, when about 20 days old. What task a bee takes on is largely governed by the level of a hormone, "juvenile hormone," in its haemolymph (the term for an insect's "blood"). This level normally increases as the bee ages, but it is also influenced by the current division of labor within its hive. Removing all of a nest's workers that are more than three days old causes some workers to become foragers within a week. These have much higher levels of juvenile hormone than either normal bees their age or fellow workers who stick to tasks within the nest. Similarly, removing all workers but foragers causes juvenile hormone to decrease in some of the youngest foragers, which then revert to working within the nest. A worker's genes also influence her hormone level and choice of tasks. A postdoctoral fellow at Cornell University, Heather Mattila, found that experimental swarms bred from queens mated with only one male will not survive a New York winter, whereas swarms bred from queens mated with several males usually do. Workers distribute themselves appropriately over the jobs to be done only if there is adequate genetic variation among them.

A honeybee colony eats about 120 kilograms of honey and 20 kilograms of pollen per year. Its foragers sometimes go 10 kilometers from the hive to find suitable flowers. Flowers appear in different places, at different times. Thus bees must be able to travel far, find suitable flowers, extract food from them, find their way back to the nest, and tell nestmates where the food is. These tasks demand the ability to move (a bee can fly 6 kilometers in 15 minutes), see and smell, find her way (bees can navigate by the sun, which requires a biological clock to tell them the time of day), and remember a "cognitive map" of its nest's surroundings. Even getting food from a flower requires some technique. A bee takes long enough to learn how to extract food from one kind of flower that, afterward, it looks for more flowers of the same species rather than learning how to handle those of other species. This "loyalty" makes them better pollinators—more likely to deliver a flower's pollen to another of the same species, enabling it to produce viable seeds.

Learning how bees tell each other where to find food, and how good it is, earned its discoverer Karl von Frisch a Nobel Prize. He found that after a forager finds a suitable patch of flowers, she flies back to the nest (fig. 3.4). After unloading her booty, the bee climbs onto a vertical comb near the nest entrance and dances—first advancing, waggling her abdomen, in a straight line along the comb, in a direction that signals the direction from the nest to the food she found. Her direction of advance makes the same angle with the vertical that the direction to the food makes with the horizontal projection of the direction from the nest to the sun. How long her "waggle run" lasts tells how far it is to the food source. The bee returns in a semicircle to repeat the waggle run, repeating this cycle for a time proportional to the ratio of her booty's energy content to the energy she used to fetch it, return to the hive, and find a food storer bee to accept the load. A dance with more waggle runs attracts more foragers to the food source, but even a few waggle runs attracts some foragers to it. The dancer makes fewer waggle runs if it took a long time to find a food storer bee, which happens when the hive's food intake exceeds its needs. Thus waggle dances not only direct foragers to the best food sources, but also adjust the hive's total intake of food to accord with its needs.

Between May and July, honeybee colonies split. The old queen leaves

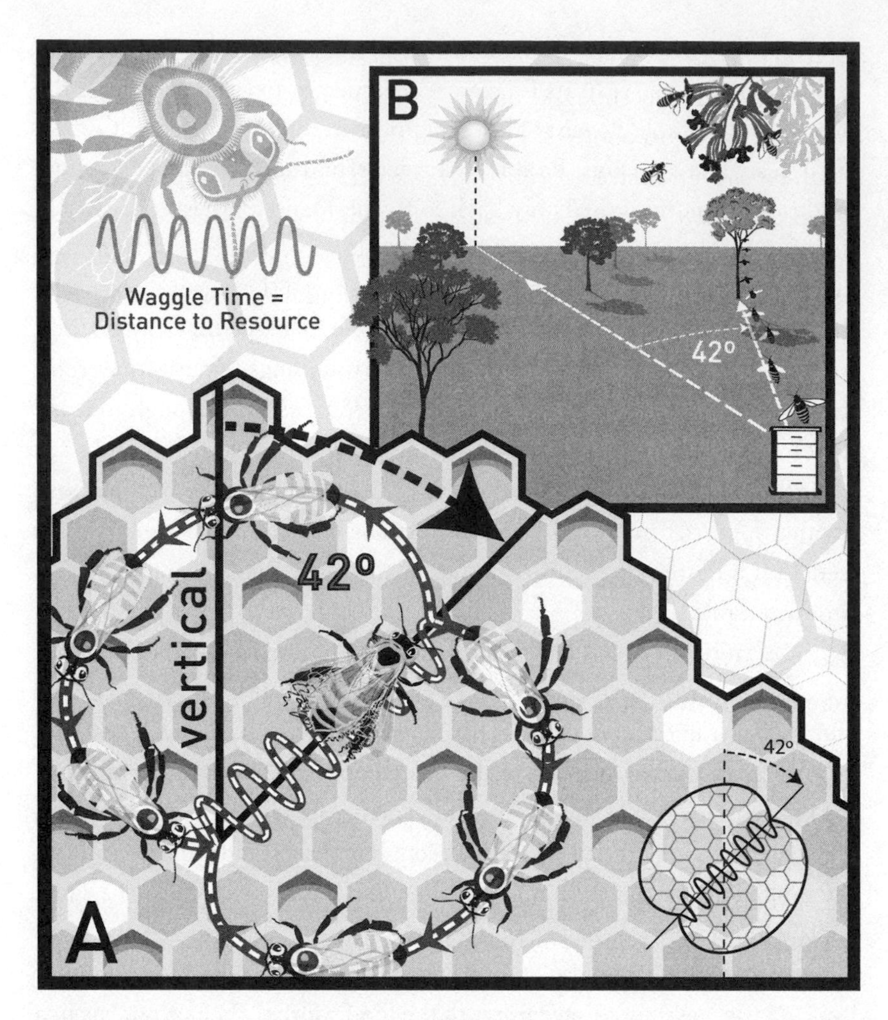

Fig. 3.4. In the "waggle dance," a honeybee signals the direction to food she has found, its distance from the hive, and its quality. Inside the hive (A), the angle of the forager's path with the vertical matches the horizontal angle of the direction to the food relative to the direction to the sun's azimuth (B). The duration of the "waggle run" (wiggly line) reflects the distance to the food; how often the bee repeats the cycle indicates its quality. (Diagram by Damond Kyllo)

with 10,000 or more workers, and new queens emerge in the old nest. If the old nest still has enough workers, a new queen may leave with an after-swarm of up to 10,000 workers, while the remaining queens fight unto death for the old nest. Each emigrant swarm must find a suitable nest site. In upstate New York, a good nest site is a south-facing tree cavity with a volume between 20,000 and 60,000 cubic centimeters (20 to 60 liters, or 5 to 16 gallons), with one opening, about 15–30 square centimeters, too small to let in much cold air, near the cavity's bottom but several meters above the forest floor. This cavity must hold the 30 kilograms of honey needed to see the nest through the winter. Good nest sites are hard to find. Learning how a swarm chooses its new home was perhaps the favorite discovery of Thomas Seeley, a leading student of honeybees, now a professor at Cornell University.

A new swarm flies about 50 meters from its former home to a nearby tree, where the bees hang together like a great beard from a branch. About 500 scouts, all older foragers, search for nest sites, looking for dark openings in tree trunks. A scout that finds a tree cavity walks around it, judging its volume by the time needed to circumnavigate it. After returning, she does a waggle dance on the swarm's surface, indicating the nest site's location and quality as if it were a patch of flowers. Then the scout returns to the site, inspects it again, returns, and dances again, averaging 15 fewer waggle runs than before. She continues this cycle, dancing only once for a mediocre site but six times or more for a good one. Each successive dance for the site has fewer waggle runs, regardless of how many scouts are dancing for other sites. Other scouts are more attracted to a nest site with more waggle runs in previous visitors' dances. When a nest site has attracted 10 or 15 scouts, dancing for that site is vigorous enough for scouts to start making a noise called "piping," which tells the swarm to warm up and make ready to fly. Usually, when the swarm is ready to lift off, nearly all dances are for the chosen site. Experiments on an island where the only sites were artificial boxes showed that swarms usually chose the best available. The way they choose compares various sites, considers all views presented, limits the time of debate, and strikes a reasonable balance between speed and accuracy of judgment. This process is one of the many signalling systems that coordinate the activities of a colony's members.

In sum, bees have evolved marvelously complex, self-regulating social groups. A group maintains appropriate stores of food in its hive; a departing swarm chooses the most suitable tree cavity for a new hive. As long as the queen stays healthy and keeps laying eggs, the colony is a marvel of adaptive function.

BABOON SOCIAL LIFE: WHAT THEY KNOW VERSUS WHAT THEY SAY

Baboons are remarkable animals. People have used them as goatherds and oxcart drivers; one was an assistant railway switchman. They live in coherent troops of about 80 animals apiece, with perhaps 7 adult males and 23 adult females. Group life has advantages: a group's members can warn each other about predators, and gang up on certain predators or competitors. A baboon has separate calls for leopards and snakes. Indeed, the sight or sound of a leopard, and a leopard alarm call, mean the same thing to a baboon: it is able to abstract at least some concepts from diverse data.

Life in a large troop also poses problems. Which baboon gets first choice of food, mates, or sleeping sites? To avoid perpetual fighting, they sort themselves out in a dominance hierarchy that governs who takes precedence over whom. A baboon's failure to know its place in the hierarchy can earn it a severe beating: consistent failure to do so can keep it from reproducing. A female who knows the troop's matrilineal kinship network—who is whose mother or maternal sister—can befriend relatives that can help protect her from being bullied or her infant from being killed by invading males. Possessing social knowledge—knowledge of the hierarchy and the matrilineal kinship network—and knowing how to benefit from it, greatly improve a baboon's reproductive success. Dorothy Cheney and Robert Seyfarth, who have studied wild baboons for many years, believe that "natural selection has led to the evolution of a mind innately predisposed to search for the patterns and rules that underlie other baboons' behavior," just as children's brains are organized for learning languages.

A baboon can recognize each of its 80 troop mates by voice or by sight. It remembers who it dominates, and who dominates it. It knows who each troop mate's mother is, and who are the offspring of any female troop mate—indeed, they are superb goatherds because they can match kids with their mothers as no person can. It infers dominance for each pair of baboons by who threat-grunts and who screams in an aggressive encounter. Remembering all these dominance relationships is a fearful job: to economize on memory, a baboon uses its knowledge of kinship, and of dominance between each pair of baboons, to infer the dominance hierarchy and matrilineal kinship network of the whole troop. This achievement amounts to forming, however unconsciously, an objective, oft-tested theory summarizing the social knowledge baboons need to succeed.

Cheney and Seyfarth demonstrated how much a baboon knows about its troop's dominance hierarchy and kinship network by playing recordings of **A** threat-grunting and **B** screaming and vice versa. If **A** normally dominates **B**, a recording of **A** threat-grunting and **B** screaming leaves a baboon unmoved, whereas a recording of **B** threat-grunting and **A** screaming shocks it to attention, especially if **A** and **B** belong to different matrilines. These recordings also suggest that a baboon's thought has syntax, because it does the equivalent of distinguishing the subject and object of a verb. It combines discrete concepts (the identities of different troop mates, dominance versus submission, type of relatedness) in the ordered structure of a rudimentary language. Do baboons innately think in syntax, as children seek to organize their words into sentences with a subject (a noun or a phrase serving as one), a verb, and an object (another noun or "noun phrase")? Did this mode of organizing thought enable baboons to infer their troop's dominance hierarchy—and track changes in it?

Although baboons' mode of thought may have enabled them to infer an objective theory, and may share the template that organized the earliest human language, baboons have little idea of what other baboons know or believe. Adult baboons, for example, will continue making alarm calls long after all troop members have heard them, and they will swim across a wide river, leaving their young, not realizing that they need help to get across. Although a baboon can assess another baboon's motives or intentions, it has no capacity for empathy—no ability to understand another baboon's feelings or to predict its behavior by imagining itself "in the other's shoes." Moreover, a baboon has only 14 distinct vocalizations, which it never combines; it cannot say all it knows. To become de facto lords of creation, human beings had to speak as well as think.

CHALLENGES OF SOCIAL LIFE: KEEPING DISEASE AT BAY

Before 1800, immigrants from the countryside maintained the populations of the world's disease-ridden cities, where more sickened and died than were born. Complex, populous insect societies face similar problems. Many ants, especially leaf-cutter ants, live in teeming nests housing a million or more—societies far more populous, and far more crowded, than the disease-ridden London or Paris of medieval times. Ants cannot depend on immigrants from healthier, less crowded settings; they are bound to the societies where they were born. How do ant societies preserve "public health"?

A leaf-cutter ant colony has one reproductive queen and perhaps a million workers, daughters of this queen. Workers cut fragments from tree leaves, and take them underground to feed a fungus (fig. 3.5). Parts of the fungus are then eaten, or fed to ant larvae. This colony is a marvel of coordinated activity. The queen produces many sizes of workers: the soldiers that protect the colony are ten times larger than the smallest workers. Each type of worker undertakes some of the colony's many tasks: locating trees with leaves their fungus can digest, cutting and bringing back leaf fragments to feed it, cultivating this fungus garden, keeping it uncontaminated, removing "used" leaf fragments and other wastes to a dump, defending the colony, caring for young, and maintaining the nest.

When the first ants, *Sphecomyrma,* evolved about 100 million years ago from solitary wasps, ancestors of the paper wasps *Vespa* and *Polistes,* they already had antibiotic-producing metapleural glands on the hind edge of their thorax, the middle segment of their bodies, as does every living ant (fig. 3.6). The antibiotic secretions of these glands play a major role in limiting microbial infection in ant colonies.

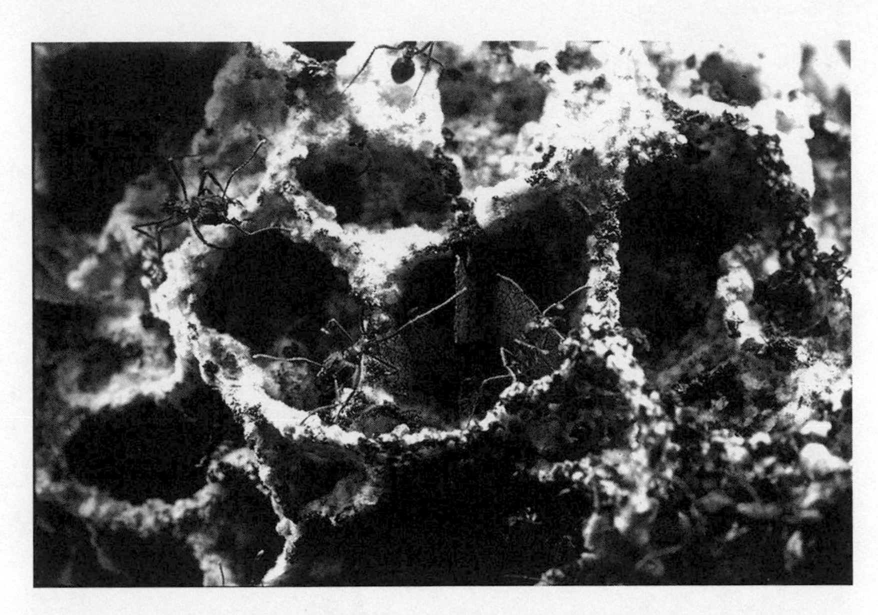

Fig. 3.5. A leaf-cutter ant colony's fungus garden, with attendant workers. The fungus turns leaf fragments, which the ants cannot eat, into fungal body parts, which the ants and their larvae eat. Mutualism between leaf-cutter ants and their fungus is maintained because the fungus reproduces only when a new queen takes a bite of her mother's fungus to start her own fungus garden. By passing the fungus on only from mother to daughter, as insects transmit gut symbionts only from parent to offspring, the symbiont's reproduction is made to depend on its host's reproductive success. (Photograph by Christian Ziegler)

Over 50 million years ago, ants evolved that lived by cultivating a fungus, parts of which they ate. Different species maintained their fungus on different types of food—dead vegetable matter, insect bodies, insect feces, and the like. Some have called this behavior the ants' invention of agriculture.

When leaf-cutter ants evolved, however, their queens, unlike the queens of the fungus-growers from which they descended, mated with many males during their one nuptial flight, to create the genetic variation that helps insect colonies survive the onslaughts of disease. Leaf-cutter ants' metapleural glands, manufactories of antimicrobial compounds, are also larger relative to body size than those of more primitive fungus-growing

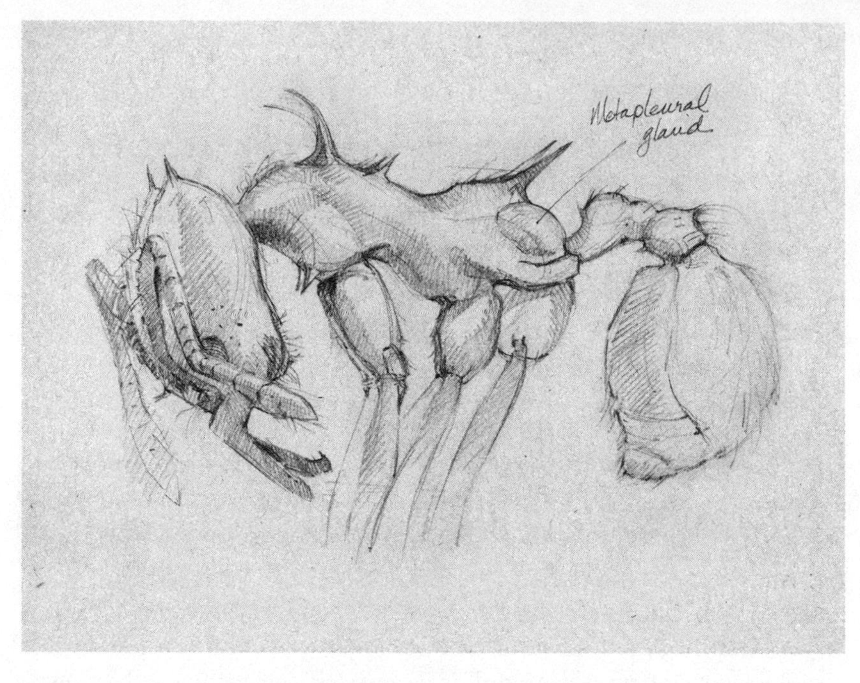

Fig. 3.6. Group life creates health hazards, which ants are adapted to minimize. A leaf-cutter ant, *Acromyrmex,* showing metapleural gland, which it uses to disinfect itself and its surroundings. (Drawing by Debby Cotter Kaspari)

ants, even those living in large colonies. It appears that leaf-cutter ants face special problems in maintaining the health of their societies. What other means do they use to preserve their "public health"?

Leaf-cutter ants, like their fungus-growing ancestors, must not only preserve themselves from disease: they must also preserve their miracle leaf-digesting fungus from contaminants that might damage or kill it (fig. 3.7). In all fungus-growing ants, the fungus gardens are vulnerable to infection by a fungal parasite, *Escovopsis.* The largest leaf-cutter ant colonies, those of *Atta,* rely on the antimicrobial secretions of their metapleural glands to control *Escovopsis.* Leaf-cutters also take great care to remove foreign particles from their fungus gardens. They discard used leaf fragments, ant corpses, and the like in dumps, normally underground. *Atta colombica,* a common species of leaf-cutter ant in central Panama, discard their refuse

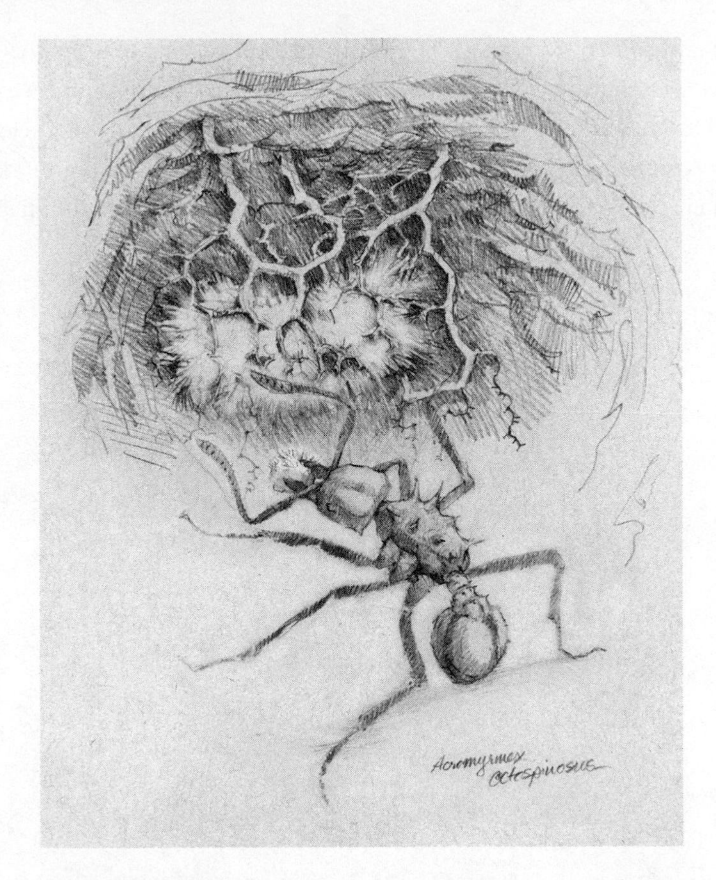

Fig. 3.7. Beginning colony life in a healthy way: An incipient *Acromyrmex octospinosus* nest, with her fungus garden suspended from rootlets on the nest cave's ceiling to avoid contamination by soil microbes. (Drawing by Debby Cotter Kaspari)

in aboveground piles downhill from the nest (plate 3.7). A special task force of workers maintains these dumps. Other colony members avoid both the dump and the workers that tend it, in order to avoid infection, so gardeners protect plants from leaf-cutters by surrounding them with refuse from their dumps. Until rains wash this refuse away, leaf-cutters will not cross such "zones of infection."

This chapter has discussed how to recognize adaptation, and provided examples of its precision and complexity in both individual organisms and animal societies. How could such precise adaptation arise? To answer, we show how living organisms derive from a common ancestry and ask: How might self-replicating organisms have arisen that natural selection could then adapt to their ways of life? How did the adaptation of organisms enable the evolution of complex, diverse, natural communities? To find out, read on.

FOUR

Life's Common Ancestry, and Its Origin

DID NATURAL SELECTION DRIVE life's diversification? If so, did modern populations descend, with divergent modifications spread by natural selection, from the same common ancestry? If so, how did life begin?

Evidence for Evolutionary Divergence from Shared Ancestors

DARWIN'S EVIDENCE FOR SHARED ANCESTRY

Several kinds of evidence convinced Darwin of evolution. First, skeletons of most land vertebrates—lizards and turtles; dinosaurs, birds, and bats; elephants, monkeys, and ourselves—look like modified versions of a basic skeleton, with a skull, jaws and teeth, a backbone of vertebrae, a tail, and two pairs of legs with knee and ankle joints, and feet with five toes, as in a toad (fig. 4.1). A mole's forefoot, a seal's flipper, and the wing of bird, bat, or pterosaur are far more similar, down to the parts they are made of, than the tools human beings make to dig, swim, or fly. Darwin inferred that vertebrates ranging from swimming reptilian ichthyosaurs and mammalian whales, flying birds, bats, and reptilian pterosaurs, running dogs, and galloping horses to slithering snakes, burrowing moles, and ourselves all diverged from four-footed ancestors by a process, natural selection, which, like a tinker, only modified and elaborated what was available.

As would be expected if Darwin was right, formerly missing early stages of divergence between groups that are now very different—the "missing links" long advertised by evolution deniers—have, one by one, turned up. Vertebrates like snakes and whales that lack arms or legs have fossil

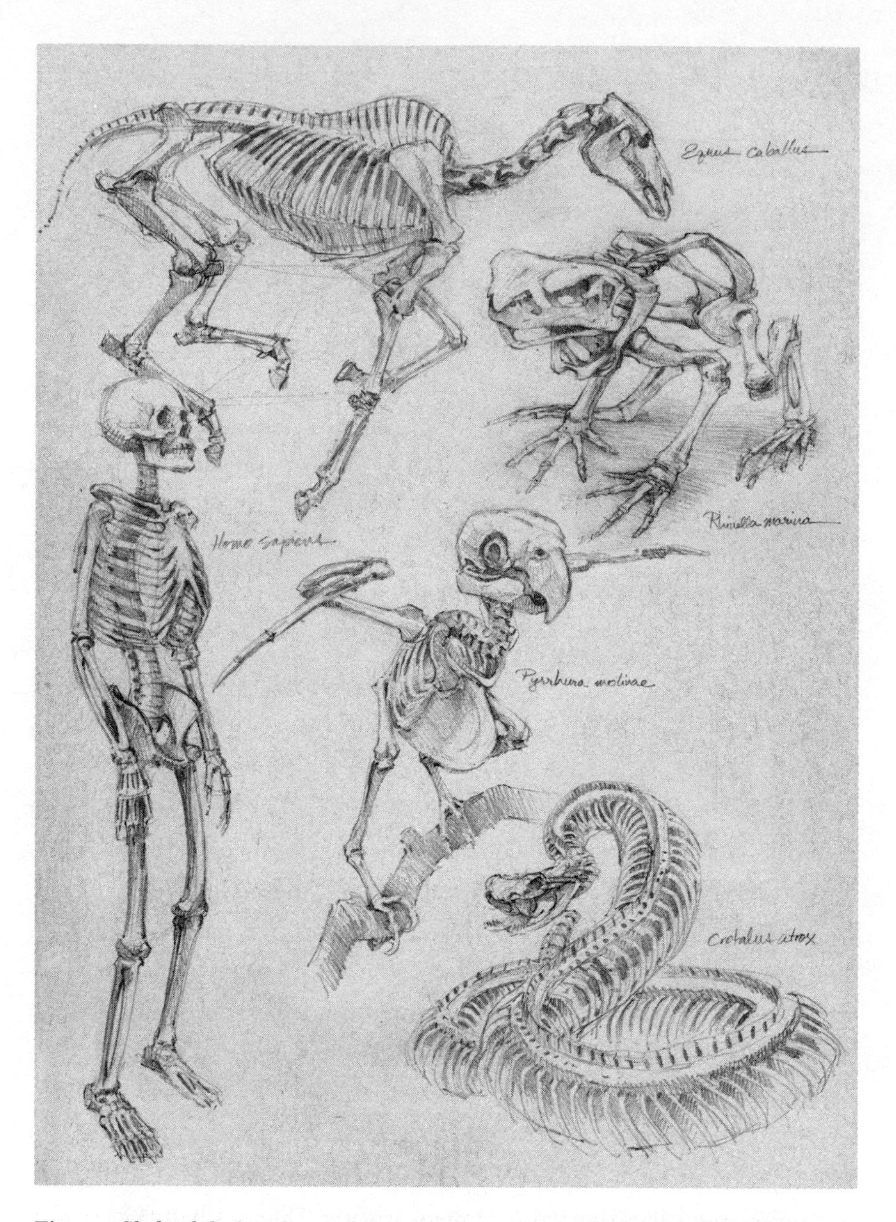

Fig. 4.1. Skeletal designs bespeaking common ancestry: Skeletons of horse (*Equus*), cane toad (*Rhinella*), human being (*Homo*), parakeet (*Pyrrhura*), and diamondback rattlesnake (*Crotalus*). These skeletons all have backbones, and heads with jaws; most have ribs and four limbs. (Drawing by Debby Cotter Kaspari)

ancestors with them. Intermediates now link birds with reptiles, and monkeys with human beings. Fossils document how amphibians diverged from fish (fig. 4.2), and whales from terrestrial ancestors of hippopotami. Of all the fossil fishes we know, *Tiktaalik* was best at walking on the bottoms of pools too shallow to swim in. It lived 375 million years ago along the shores of rivers meandering through subtropical forests in what is now Ellesmere Island in the Canadian Arctic. The fleshy stumps of their lobe fins, like those of coelacanths and lungfishes, had both fish-style fin rays and the skeletal beginnings of fore and hind feet, with primordial wrists, ankles, and arm, leg, finger, and toe bones. These beginnings of feet, a pelvis, and shoulder blades enabled *Tiktaalik*'s descendants to colonize the land. By doing so, *Tiktaalik*'s descendants committed their progeny to a future full of opportunity.

Fossils also show how whales evolved from land mammals (fig. 4.3). The fossil skull and teeth of the earliest known whale, *Himalayacetus* (*Cetus* is Latin for whale), which lived in India 52 million years ago, suggest that it ate fish in shallow water, both fresh and salt. Its ear bones were not yet specialized in any way for hearing underwater. The 49-million-year-old fossil of another fish-eating whale, *Pakicetus* from Pakistan, had ear bones indicating some ability to hear both in air and underwater. When underwater, however, it could not tell where sound was coming from. A *Pakicetus* skull was found with fossil land mammals; *Pakicetus* must have spent some of its time on land. Another Pakistani whale with hands and feet was the 47.5-million-year-old *Ambulocetus*. This whale was as big as a 300-kilogram male sea lion—twice the size of *Pakicetus*. It could walk on land and swim in the sea, powered by its hind feet and the up-and-down motion of its tail. About 46 million years ago, a later Pakistani whale, *Rodhocetus,* was more committed to sea life. It had larger hands and feet than *Ambulocetus,* but its hind legs were a third shorter, giving it a more streamlined form. *Rodhocetus* was perhaps the first whale to hunt in deep water. The back half of its spinal column anchored powerful muscles. This whale swam by paddling with its wide hind feet and undulating its powerful tail. Soon after, *Protocetus,* with an even more powerful tail, fished in deep water off Egypt. Its pelvis was no longer attached to its backbone: it could not walk on land. Thirty-five

Fig. 4.2. Salamanders inherited many features from lobe-finned fish that bear witness to their aquatic ancestry, as shown in the forelimbs and skeletons of coelacanth (*Latimeria*) and *Tiktaalik,* with incipient limb, hand, and finger bones inside the fins, and a salamander with its forefoot skeleton. (Drawing by Debby Cotter Kaspari)

Fig. 4.3. How whales evolved from forest-dwelling browsers: Skeletons of a modern lesser Malay chevrotain (*Tragulus javanica*), similar to *Indohyus*, the terrestrial ancestor of whales; *Pakicetus*, an early whale that could still walk, and *Ambulocetus*, which could walk no better than a modern seal, from the Eocene; and a modern humpback whale (*Megaptera*). (Drawing by Debby Cotter Kaspari)

million years ago, 16-meter whales, *Basilosaurus,* swam off the Egyptian shore with their great horizontal fluke: their remnant hind limbs and pelvis were useless. Judging by the structure of their ear bones, they could hear underwater nearly as well as their modern counterparts, some of whom are famous for their underwater songs.

Darwin also knew that early stages of embryos of different vertebrates are far more similar than their adults (fig. 4.4). Like fish, young human embryos have tails and gill slits. A fetal whale has rudimentary feet, like land mammals and unlike today's adult whales. These resemblances also reflect descent with divergent modification from common ancestors. Selection is strongest after parental care ceases and young must make their own livings, so adults of different species differ far more than their embryos.

Finally, as Darwin realized, if all living things descended with modification from common ancestors, they must have spread from their ancestors' homes to where they live now. Plants and animals of oceanic islands never connected to the mainland, such as the Galápagos, 600 kilometers west of Ecuador, the Cape Verde Islands, 500 kilometers west of West Africa, or the mid-Pacific Hawaiian Islands, must have had mainland ancestors that crossed wide expanses of ocean. The Galápagos and the Cape Verdes are volcanic islands with dry lowlands, quite unlike the nearest mainland. Yet Darwin found that most species living on the Galápagos, and on the Cape Verdes, either also live in the nearest continent, or have close relatives there. Those denizens of the nearest continent best at crossing water must have colonized these islands. Thus some birds and bats, but no walking mammals, reached the Hawaiian Islands. Some insects, and spiderlings borne by silken threads, were blown across, and snails crossed on the feet of birds. Earthworms or frogs are more easily killed by salt water, and less easily carried by birds; no earthworm or frog reached the Hawaiian Islands, or even the Galápagos. Some weeds whose seeds were light enough to be blown across, or which could stick to bird feathers or survive in their guts, but few large-seeded trees, reached Hawaii. Indeed, the closest mainland relatives of the trees and other large-seeded plants native to islands like those of Hawaii are usually (but not always!) small-seeded weeds or herbs. Darwin asked: If each species was separately created for its own habitat,

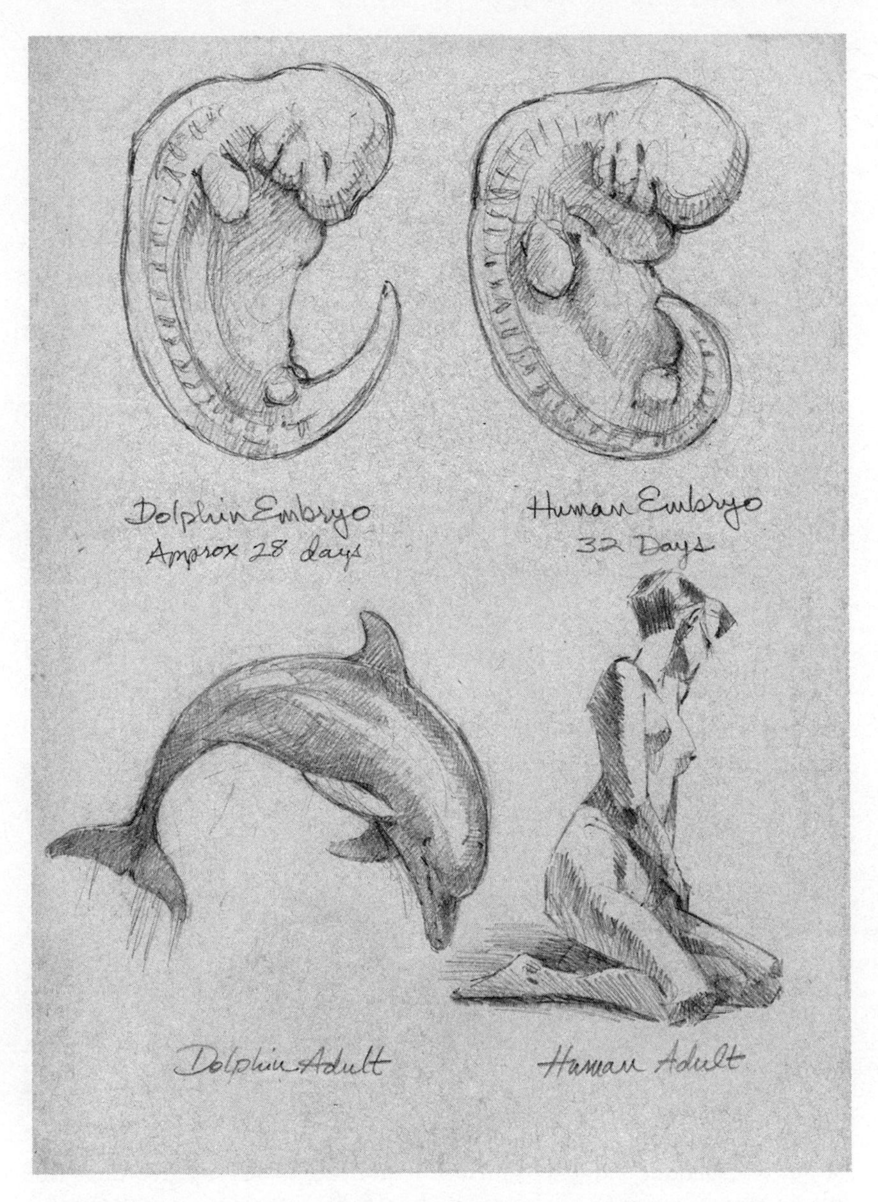

Fig. 4.4. Embryo and adult of dolphins and human beings. Embryonic mammals resemble one another far more than adults. (Drawing by Debby Cotter Kaspari)

as creationists infer from the Bible, why do most island species either have their closest relatives on nearby mainland, or belong to groups whose mainland members could disperse far over water?

BIOCHEMISTRY, MOLECULAR BIOLOGY, AND LIFE'S COMMON ANCESTRY

The chemistry of life also reflects descent with modification from common ancestors. James Watson, co-discoverer of the double helix structure of deoxyribonucleic acid (DNA), observed: "The basic biochemical reactions upon which cell growth and division depend are the same, or very similar, in all cells, those of microorganisms as well as of higher plants and animals." Moreover, all living beings transmit heritable characteristics to their offspring by a recipe encoded in DNA—a program for transforming a fertilized egg into a functional adult. The DNA in genes, which are arrayed in chromosomes, program an ordered set of chemical reactions, each catalyzed by a specific enzyme. Amylase, for example, is an enzyme in our saliva that catalyzes the breakdown of starches into their component sugars. These reactions build, power, and maintain the organism. Each enzyme's structure (shown later in fig. 7.4) is programmed in a code expressed as a sequence of four "letters"—the nucleotides adenine, thymine, guanine, and cytosine—in a DNA molecule. A sequence of three nucleotides—a codon—programs either an amino acid or the start or stop of an amino acid sequence. The codon sequence for an enzyme programs a sequence of amino acids (nitrogen-containing molecules) which, when formed, folds up in a specific way to make a protein, often, but not always, an enzyme.

Proteins of most living things from bacterium to human being are made from the same twenty amino acids. Living beings all translate a DNA strand's nucleotide sequence into a protein in the same way, first transcribing the DNA sequence onto a complementary sequence of messenger RNA. RNA, ribonucleic acid, is a molecular strand like DNA, but less stable because thymine is replaced by a similar base, uracil, which also pairs with adenine. The messenger RNA causes a protein to form at a structure called a ribosome where each codon attracts a specific molecule of "transfer RNA," which in turn attracts the corresponding amino acid and attaches it to the growing protein, which then folds into a functional enzyme. Such enzymes are manufactured only when needed. For example, a repressor protein normally blocks synthesis of an enzyme that breaks lactose into usable sugars. When lactose is present, however, the repressor combines with it, allowing manufacture of this now useful enzyme. Enzymes are also designed to coordinate their reactions with others. Some enzymes function only when combined with a molecule that is the starting point for a different series of reactions, thus coordinating these reactions. Others function only when not combined with the end product of their own reaction sequences, which prevents overproduction of that product. In all organisms, the genetic code, how it is translated, and how the reactions these genes program are controlled and coordinated, are identical, reflecting life's common ancestry.

How Life Began

When living beings—entities using resources from their habitats to replicate themselves—first appeared, the natural selection that resulted began to transform the earth. How could materials and processes devoid of purpose give rise to beings organized to harness energy to live and reproduce? Obviously, no human being saw this happen. Nevertheless, some guidelines help us to think about, and recognize clues to, where and how life began. First, chemical reactions that constantly provided energy which could be harnessed to do the work of growing and reproducing, must have appeared before genes evolved to exploit, modulate, and control these reactions.

Second, reactions supplying power to do any kind of work, including the work organisms must do to live and multiply, only arise from enduring disequilibria. A water mill draws the power needed to grind corn from the disequilibrium between water above and below a cliff: corn is ground because water falling over the cliff turns the mill wheel. This disequilibrium endures because sunlight evaporates water from wet surfaces, some of which condenses as rain in the highlands, keeping the mill's waterfall flowing. The mill thus harnesses a work cycle to grind corn. Likewise, organisms must harness one or more work cycles to live, grow, and multiply. Life could not begin in a warm, still pond, even one full of life's chemical building blocks such as dissolved ammonia, phosphate, and carbon dioxide. Such ponds lack the steady power supply needed for these compounds to assemble "by accident" into organisms that could live and reproduce.

Third, an event as many-faceted as the origin of life, which involves harnessing energy, forming distinct individuals, and replicating their kind could only happen in a setting where it could develop one aspect, one piece, at a time.

When life began, nearly four billion years ago, water covered the whole earth, except for a few volcanic islands. Asteroids, far larger than the bolide 10 kilometers (6 miles) wide that hit the Yucatán 65 million years ago, struck the earth from time to time, boiling the oceans. Atmosphere and oceans lacked free oxygen, so no ozone layer shielded the earth from ultraviolet radiation, making life at the surface impossible. The atmosphere was largely carbon dioxide. Where could life begin in this hostile world?

DEEP-SEA VENTS: SUBSIDIES THAT MADE LIFE POSSIBLE?

Deep-sea vents were discovered in 1977 atop a mid-ocean ridge where magma wells up volcanically from below the sea floor and spreads out to both sides. Although the vented fluid is far too hot (350° C) for life to begin there, some biologists wondered whether life first developed where cooler vented fluids met seabottom water. Ridge-top vents, called "black smokers," spew out acid black water, superheated by magma and enriched with dissolved iron and hydrogen sulphide. Bacteria around these vents can exploit the disequilibrium between vented fluids and oxygen-rich seabottom water by catalyzing the energy-releasing reaction between hydrogen sulphide and the oxygen dissolved in seawater to form water and sulphur. In turn, the swarms of bacteria living by this and related reactions support a diverse multitude of animals.

Talk about equilibria is a bit like Aristotle's physics, where objects "prefer" certain states. Water falls as low as it can get. Similarly, hydrogen atoms "prefer" joining oxygen to joining sulphur atoms: when they meet, hydrogen sulphide reacts with dissolved oxygen to form water and sulphur, but water and sulphur never combine to make hydrogen sulphide and oxygen. Like any other "spontaneous" chemical reaction, the reaction between hydrogen sulphide and oxygen releases energy.

A biochemist, Michael Russell, guessed that life began in a cooler type of sea-floor vent, away from mid-ocean ridges. In 1981, 350-million-year-old chambers of what appeared to be an off-ridge deep-sea vent, lined with masses of cell-sized bubbles, were found in an Irish silver mine. These bubbles seemed to be suitable sites for life's chemical reactions to develop. His son's chemistry set showed him how these bubbles could form. Pouring a solution of cobalt chloride into an alkaline solution of sodium silicate creates the many-colored spires of a "chemical garden." Russell's son discovered that his "garden's" spires were hollow and full of bubbles. Inspired by this discovery, Russell poured an alkaline solution of sodium sulphide into a mimic anoxic ocean containing dissolved iron chloride. As expected, he got chemical gardens of hollow spires, many filled with masses of microscopic bubbles. These bubbles had porous walls of iron sulphide (FeS); small molecules passed through their walls more easily than large.

Modern vents resembling Russell's fossil were found in 2000 (fig. 4.5), about a kilometer deep, 16 kilometers from the mid-Atlantic ridge. There, magma spreading from the mid-ocean ridge had cracked the earth's crust, allowing seawater to react with the mantle below in a way that generated hydrogen. When this enriched, hot (90° C), alkaline water spewed out through such "alkaline vents," chemicals in the water reacted with other chemicals in the ocean to form many-chambered white towers up to 60 meters (200 feet) tall. These towers had walls of limestone, clay, and fool's gold (iron pyrite, FeS_2), riddled by channels and pores. Some chambers were filled with interconnected bacterium-sized bubbles with thin walls of calcium carbonate. In homage to Plato's "Lost City of Atlantis," these alkaline vents were called the Lost City Hydrothermal Field. Thanks to the disequilibrium between hydrogen from the vents and carbon dioxide in the ocean, alkaline vents also support multitudes of microbes. Microbes of the two earliest divisions, bacteria and archaea, derive both energy and organic compounds by combining hydrogen with carbon dioxide; in this

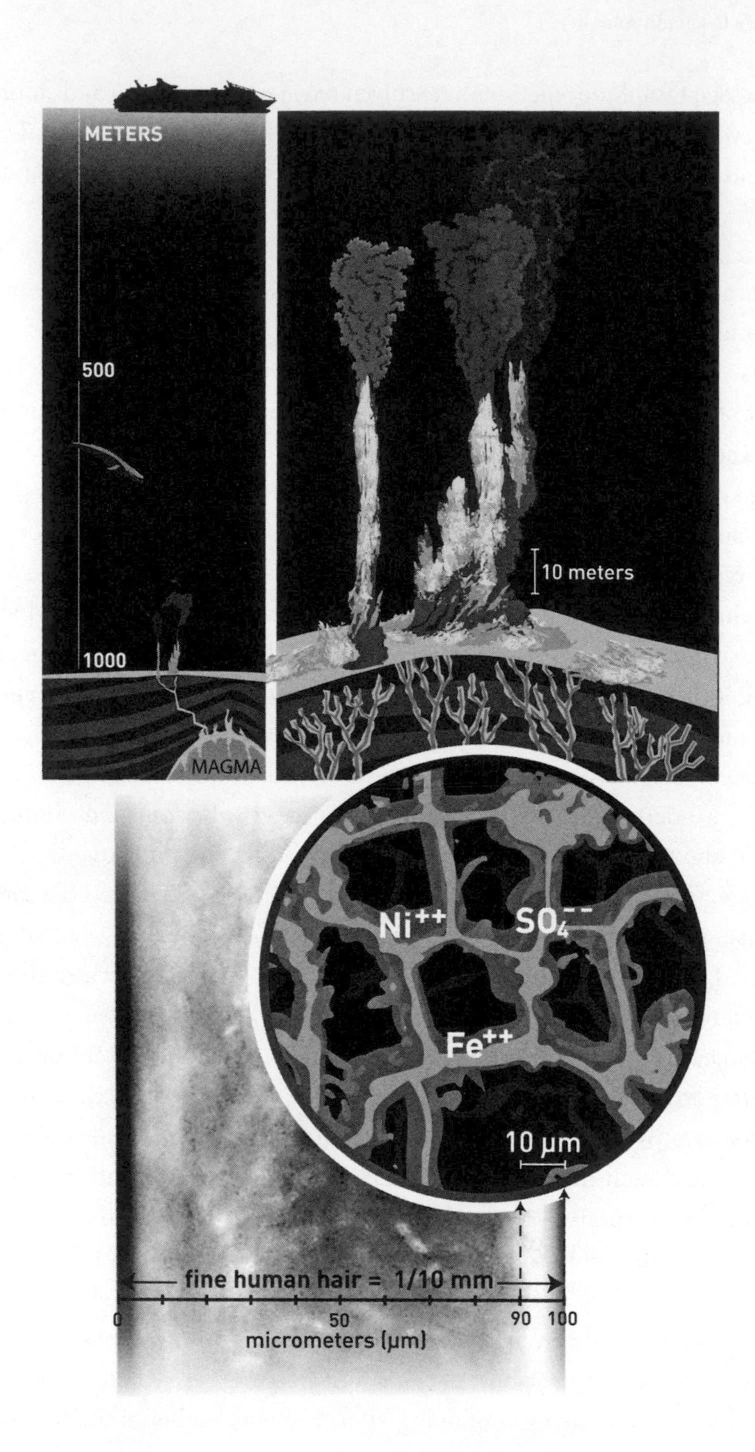
METERS
500
1000
MAGMA
10 meters
Ni^{++}
SO_4^{--}
Fe^{++}
10 μm
fine human hair = 1/10 mm
0
50
90
100
micrometers (μm)

process, bacteria produce vinegar (acetic acid); archaea, methane. Abiotic reactions in the vent involving hydrogen and carbon dioxide also produce some methane; under it, others that also involve ammonia produce amino acids, the building blocks of proteins. Were such vents the flutes through which was played the overture to the Dance of Life?

HOW DID "PROTOLIFE" ACQUIRE AND DEPLOY ENERGY?

Living things must both procure energy and use it appropriately. To do this, they need a steady supply of energy, catalysts to facilitate appropriate chemical reactions, and a steady supply of cell-sized compartments that confine products of these reactions, allowing them to combine and form new products. Alkaline vents supply all three.

First, the fluid these vents exhale provides a steady supply of hydrogen, which, mixed with carbon dioxide from the surrounding seawater, can release energy. Today, combining hydrogen stripped from water molecules with carbon dioxide from the air supplies the final stage of photosynthesis.

Second, in those oxygenless days, combination of sulphide in the vented fluid with iron dissolved in the seawater above would have filled these vents with a steadily renewed supply of cell-sized bubbles with porous iron sulphide walls, which smaller molecules could pass through.

Third, hydrogen will not react spontaneously with carbon dioxide. As procuring energy from a high crater lake requires digging a tunnel through the volcano wall to let the water fall, so hydrogen and carbon dioxide must be induced to undergo a series of chemical reactions before energy is released or organic compounds form; these reactions must be cat-

---

Fig. 4.5. (*opposite*) An alkaline vent. Unlike chimneys of black smokers, which last 50 years, an alkaline vent's chimneys last over a hundred thousand years, allowing time for the processes of life to get a start. An alkaline vent (upper panels) provided a template for cellular pre-life's metabolism, which used chemical disequilibrium to subsidize the metabolism and self-replication allowing life to evolve. Alkaline vents had microscopic "prefabricated" cells with iron sulphide walls (bottom). Did they house pre-organisms? (Artwork by Damond Kyllo)

alyzed. The ocean then contained dissolved nickel, so bubble walls became studded with compounds of iron, nickel, and sulphur that catalyzed the reactions yielding, however slowly and crudely, energy and the various building blocks of life, such as sugars, amino acids, and nucleotides, from hydrogen, sulphide, and ammonia in the venting fluid and carbon dioxide and nitrate in the seawater. Reactions that must be catalyzed can be controlled. Moreover, the chemicals thus catalyzed must have included better catalysts. In sum, alkaline vents provided a reliable source of energy, the means for controlling its release and directing its use, and cell-sized bubbles, and confined spaces where organic compounds could react with each other rather than diffuse away.

The "pathway" by which catalysts on bubble walls produced energy and organic compounds from hydrogen and carbon dioxide was an abiotic precursor of the acetyl coenzyme A pathway—the metabolic pathway that archaea and bacteria in modern alkaline vents use for this purpose. Modern organisms use five pathways to make organic compounds from carbon dioxide; the acetyl coenzyme A pathway is the only one that also produces energy rather than needing it to power its work. This pathway uses the simplest enzymes (protein catalysts) to control its reactions. Finally, the active sites of these enzymes, sites where reacting molecules are actually juxtaposed, are compounds of iron, nickel, and sulphur very like those studding bubble walls in ancient alkaline vents. Are these footprints of this pathway's origin in alkaline vents?

Iron sulphide cells were also where respiration, which generates energy for living, could evolve. In bacteria and archaea, respiration occurs only on the outer membrane. Today, oxygenic respiration combines hydrogen from sugar with oxygen to make carbon dioxide and water. This is an energy-releasing reaction analogous to fire, but bacteria control it far more precisely than we do fire. Microbes use the released energy to pump protons (hydrogen atoms stripped of their electrons, therefore positively charged) outside the cell membrane, creating a disequilibrium, a "proton motive gradient," between an acidic outside and an alkaline cell interior. This respiration-driven proton-motive gradient provides the energy that drives the assembly of all compounds bacteria and archaea need to live and reproduce. Before oxygen entered the atmosphere, as in anoxic settings today, microbes used other energy-releasing reactions to maintain this gradient. The technology most organisms use to create this gradient is astonishingly complicated. How could such complex technology have evolved? In alkaline vents, vented fluid meets more acidic ocean water. Fluid inside the iron sulphide cells was less acidic than the water outside: the difference was as great as in a modern proton motive gradient. The proton motive gradient across iron sulphide "membranes" could help drive the synthesis of organic compounds in vents. Could it also be where early stages of respiration evolved? Did crude organic membranes line the insides of bubble walls, restricting the inward flow of protons? If so, then once genes appeared, selection favored means to maintain the proton gradient, and an impressive molecular device evolved that all living beings use to derive energy from this gradient to assemble the organic compounds they need.

THE ORIGIN OF REPLICATION: NATURAL SELECTION TAKES OVER

How could these reactions become controlled in a way that allowed reproduction? DNA cannot replicate itself: its replication must be catalyzed by enzymes, which DNA must in turn encode. The answer appears to be ribonucleic acid (RNA). One of the macromolecules that life depends on, RNA consists of a single chain of chemicals called ribonucleotides. RNA sequences catalyze some reactions. Indeed, they probably entered the history of protolife as catalysts particularly apt at facilitating the formation of amino acids and joining them together to make peptides and proteins. Even now, some RNA compounds catalyze essential reactions. As some modern catalysts of ancient reactions have compounds of iron, sulphur, and nickel at their active sites, so others have active sites with bits of RNA, another footprint of how life evolved. Some of these catalytic RNA sequences, however, could also replicate themselves unaided, albeit inaccurately. The earliest replicating RNA sequences may have resembled modern RNA viruses, which catalyze or modify chemical reactions in their cell to replicate themselves. When such sequences modified chemical reactions in the vent's acetyl coenzyme A pathway to replicate themselves, natural selection began to play a role in life's evolution.

These early virus-like RNA sequences would have replicated themselves far less accurately than DNA, which had not yet evolved. To persist, the sequences had to be so short that most replicates did not contain fatal errors. No one sequence could be long enough to catalyze all the reactions needed to replicate itself, but a group of RNA sequences could replicate themselves if each happened to catalyze a chemical another needed. Such "cooperation," however, is vulnerable to cheaters, which benefit from the activities of other sequences while doing nothing to enhance their replication. A "cheating" RNA sequence might replicate faster if it did not catalyze a reaction whose products other RNA sequences in its group needed to replicate themselves. Most likely, what saved protolife from cheaters was the confinement of groups of RNA sequences to distinct bubbles in a vent. If a bubble's group of RNA sequences helped one another reproduce, the group colonized new bubbles, whereas a cheating RNA sequence that blocked its own group's reproduction would die before infecting other functional groups like a parasitic virus. Therefore, a cell's sequences would normally live or die as whole groups. If groups "reproduce" by occupying adjacent, newly formed bubbles, and if groups rarely exchange sequences, then differential replication—natural selection—favors more cooperative groups, which are more functional and reproduce more rapidly, and something worthy of the name of life begins to emerge, shaped by selection among contents of different bubbles for cooperation among a bubble's bits of RNA code.

Natural selection became far more effective when a group's RNA sequences were read off master DNA sequences, which were much more stable. The genetic code, and modern means of transcribing DNA to RNA and translating RNA sequences into proteins, evolved before archaea and bacteria diverged. In life's latest common ancestor, genes expressed themselves as they do now. First, a complementary "messenger RNA" was read off the gene's DNA sequence. The messenger diffused to a ribosome, a site of protein synthesis. The ribosome moved from triplet to triplet of the messenger. At each triplet, a "transfer RNA," with the complement to that triplet at one end and the appropriate amino acid at the other, joined with the messenger in a way that added its amino acid to the growing protein. When the protein was completed, it was released to fold into the shape that catalyzed a reaction that helped its organism live and reproduce.

Modern means of DNA replication, however, must have evolved after bacteria and archaea diverged: they use completely different enzymes to replicate their DNA. These RNA sequences must have behaved as retroviruses do today, catalyzing the formation of complementary DNA sequences to store the data coded in the RNA more reliably. Here, RNA was central to DNA replication: a complementary RNA strand was read off from a DNA strand and a copy of the original DNA strand was made from this RNA strand. Moreover, archaea build their cell walls and cell membranes from chemical compounds and enzymes utterly different from those that bacteria use for the same structures. Apparently, the ancestors of bacteria diverged from those of archaea when a "microbe" was still separated from "outside" by the wall of its iron sulphide bubble, and only afterward independently evolved cell membranes, cell walls, and DNA replication. When cells evolved that could replicate themselves, complete with cell walls, cell membranes, and DNA, the DNA could fairly be called that organism's recipe for living, its hypothesis for how to live in its environment and make other organisms like itself.

In short, alkaline vents may have provided sites where the seemingly irreducibly complex problem of the origin of life could be broken down into parts that could be solved sequentially. Moreover, each part could be solved in stages. Thus the origin and refinement of metabolism probably started well before genes controlling and coordinating metabolism began to evolve. Many different virus-sized genetic elements had evolved before the coordinated replication of genomes was achieved. Confining reactions and their catalysts to cell-sized bubbles allowed the competitive process of natural selection to favor cooperative groups of RNA sequences. The evolution of cell membrane respiration, creating proton motive gradients, was well advanced before cells possessed walls and membranes of their own make. Our picture of how some of these problems were solved is still exceedingly sketchy; some of the stories involved are still educated guesses. The alkaline vent work, however, provides compelling evidence that life had a mechanistic origin, and suggests how beings capable of using resources to reproduce, which natural selection automatically endows with the purpose of reproducing themselves, could arise from processes as devoid of purpose as the downhill flow of water or the rusting of iron.

FIVE

Diversification

Competition, Innovation, and Diversification

Just as human beings spread from Africa to other continents and the Malay Archipelago, and from tropical savannas into forests, grasslands, tundra and polar ice, looking for places to live, so living beings slowly spread to nearly all of the earth's habitats. As descendants of hunter-gatherers devised new ways to make livings, starting with medicine, tool-making, trade, agriculture, and pastoralism, so descendants of the first living beings created new ways of life, each providing openings for others. The French novelist and essayist Georges Bernanos observed that "each civilization is the habitat of generations of human beings who have shaped it for their use." Likewise, the story of life is one of living beings slowly turning this planet into an ever more suitable home for life. Just as human beings domesticated nature to create their civilizations, so other living things transformed their surroundings, shaping the earth's habitats, and their chemistry, in ways that increased life's diversity and production of edible matter. They did so without foresight, as new, better adapted variants automatically replaced their predecessors.

EARLY MICROBIAL LIFE: THE CHEMICAL ACHIEVEMENT

Soon after free-living archaea and bacteria with cell walls and membranes of their own make emerged from alkaline vents, they found new ways to make livings. About 3.5 billion years ago, some bacteria used energy from light to make sugar from hydrogen sulphide and carbon dioxide. Where usable nitrogen was scarce, others used energy from sugar to make the ammonia needed to manufacture proteins from atmospheric nitrogen. Yet others combined nitrate with sugar to release energy, keeping the oceans from becoming nitrate brines. The balance of self-interest among these bacteria made the earth's oceans a better home for life.

Nearly all living beings on earth are guests of its cyanobacteria and green plants (plate 5.1). These "primary producers" live by "oxygen photosynthesis," in which a cell's chlorophyll traps photons, elementary particles of light, and uses their energy to make sugar and oxygen from carbon dioxide and water. Oxygen photosynthesis evolved just once, in cyanobacteria, about three billion years ago. This crucial achievement did for the natural world what agriculture did for humanity.

Oxygen photosynthesis had another decisive effect: enriching the atmosphere with oxygen. No lifeless world's atmosphere has oxygen. More oxygen led to more life, in two ways. First, ultraviolet light turned some of the upper atmosphere's oxygen into an ozone layer, shielding the earth's surface from lethal doses of this radiation, which made it easier to live on land. Second, oxygen made possible aerobic respiration, a carefully regulated process of combustion, the energy-releasing reverse of the photon-driven reaction of photosynthesis. Aerobic respiration releases more energy from food than any form of anaerobic respiration. Stocking the atmosphere with oxygen, however, took time. Oxygen photosynthesis evolved three billion years ago, but oxygen appeared in the atmosphere only about 2.4 billion years ago. There was enough oxygen 100 million years later to form an ozone layer, blocking most ultraviolet light. Air bubbles sealed in ancient salt deposits showed that, 815 million years ago (and probably far earlier) atmospheric oxygen was half today's level. Until 700 million years ago, ocean water below the surface lacked oxygen, because cyanobacteria dominated primary production, and decomposition (oxidation) of their minute, slowly sinking bodies drained the oceans of oxygen. Only when sponges began to spread, filtering out cyanobacteria and turning them into turds, often buried undecomposed, did the oceans begin to oxygenate.

Diversification and Evolutionary Progress

Evolution involves progress. Since the Precambrian, animals of many groups evolved and coordinated new sensory, locomotory, digestive, navigational, cognitive, and social skills to exploit new foods, catch more elusive or powerful prey, escape predators better, or migrate to where food is seasonally abundant. Animals evolved that cooperated with plants, or with one another, in mutualisms, some crucial to their ecosystem's productivity. As J. B. S. Haldane and Julian Huxley wrote in 1927, evolution raises "the upper level of organization reached by living matter, while still permitting the lower types of organization to survive. This . . . gradual rise in the upper level of control and independence . . . in living things . . . may be called *evolutionary* or *biological progress.*"

MAKING COMPLEX, HIGHER ORGANISMS: TURNING BACTERIA INTO LIVE-IN POWER PLANTS

The first step in evolving plants, animals, and fungi from microbes was taken about two billion years ago. An archaean of a newly discovered kind, the Lokiarchaeota, which apparently can engulf other cells, engulfed or was invaded by bacteria that could respire aerobically *and* multiply within their hosts. These archaeans are the ones most closely related to eukaryotes (organisms whose chromosomes are in a nucleus enclosed by a membrane). Engulfing prey, which otherwise only eukaryotes can do, made the archaean host good at procuring food; aerobic respiration made their live-in bacteria good at extracting energy from it. These complementary abilities favored tight mutualism between these archaea and their live-in bacteria. Mitochondria, the oxygen-respiring power plants of eukaryotic cells, evolved from these bacteria. Mitochondria are so perfect for their role that until recently biologists refused to believe that they descended from independent organisms. Yet mitochondria divide, like cells, and they have their own genes, whose DNA reveals their bacterial ancestry. Eukaryotes evolved by combining *and* modifying preexisting organisms that began to live in symbiosis.

The origin of eukaryotes was a unique, crucial event: all organisms with mitochondria descend from the archaean that first "tamed" their organelles' bacterial ancestors. This event opened great possibilities. Proton-motive gradients are essential in respiration, which supplies the energy to live. They occur only on a bacterium's outer membrane, its surface. Thus bacteria must be small, for as a cell grows, its volume grows faster than its surface area (when a cell's diameter doubles, its surface area quadruples and its volume grows eightfold). A large cell's volume thus outruns the energy from respiration at its surface. The surface areas of their many mitochondria let eukaryote cells escape this size limit (fig. 5.1). Larger size allows larger genomes, allowing more refined responses, better suited to local events and changes, and more flexible energy use. Thus all organisms with complex division of labor among their cells are eukaryotes.

This transition imposed many changes. The genome was reorganized into distinct chromosomes in a nucleus separated by a membrane from the rest of the cell: eukaryotes are named for their distinct nuclei. Orderly sexual reproduction evolved, where two cells fuse, homologous chromosomes pair, and the fused cells divide again into daughter cells, each with a complete genome. Precise pairing nearly blocks horizontal gene transfer, the transfer of genes between organisms so distantly related that the size, number, or gene arrangements of their chromosomes differ markedly, so each interbreeding population now develops its own pool of coevolved genes. Meiosis requires choosing mates with identically arranged genomes: making it possible to diversify into distinct species. Evolution sped up after large, active animals appeared.

In oxygen-rich nearshore habitats, eukaryotes, one-celled "protists," gave rise to ancestors of plants, animals, and fungi about 1.6 billion years ago. Over 1.2 billion years ago, a eukaryote acquired new symbionts, photosynthetic cyanobacteria, ancestors of the chloroplasts in leaves, and red and green algae, that enabled these plants and algae to photosynthesize. Ancestral kelps, diatoms, and other algae, however, acquired chloroplasts by engulfing eukaryote cells of other species that already had them.

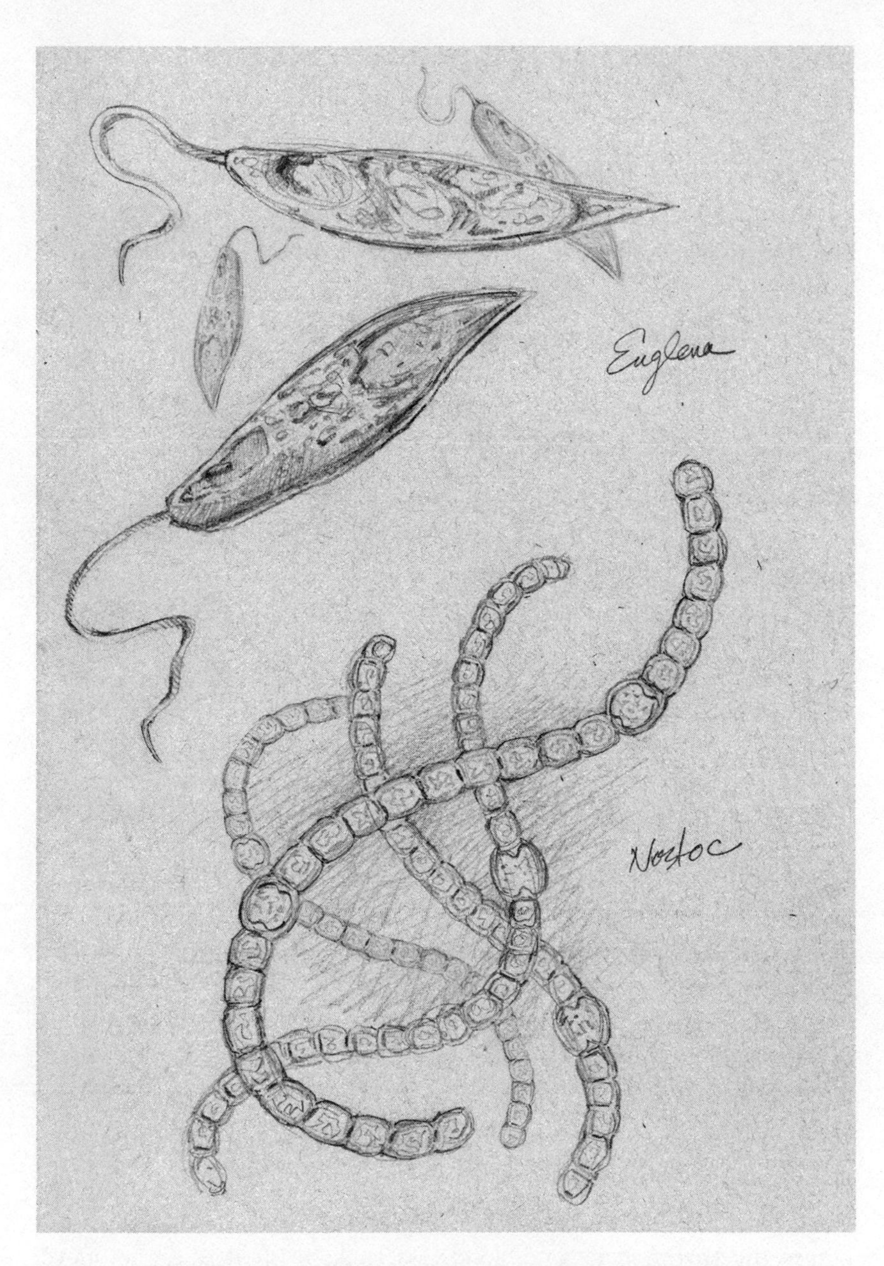

Fig. 5.1. Turning bacteria into live-in power plants gave rise to eukaryotes, which in turn gave rise to larger, more complex one-celled organisms. A eukaryote, the photosynthetic "protist" *Euglena* (0.1 mm long, top) and a cyanobacterium, *Nostoc* (cell diameter 0.004 mm, bottom). (Drawing by Debby Cotter Kaspari)

MAKING COMPLEX, HIGHER ORGANISMS: ORIGINS OF COMPLEX MANY-CELLED BEINGS

Starting about 700 million years ago, as the oceans began to oxygenate, another major transition occurred several times, transforming clumps of clonally produced cells, each descended from a sexually produced, genetically unique fertilized egg, or "zygote," into multicellular individuals. For this to happen, cells had to distinguish clone-mates from outsiders, exclude the outsiders, stick to and feed each other, and coordinate their activities by chemical signals between cells. These transitions all involved cooperative division of labor among a clump's cells, which eventually entailed differences in these cells' structure as well as their behavior. The first was between reproductive cells and cells with other functions. Plants descend from one such transition, animals from another, kelps and related seaweeds from a third, fungi from two others. Multicellular forms of life range from sea lettuce, *Ulva,* and trees to jellyfish, oysters, and people. Multicellular organisms that function as genuine individuals are all composed of eukaryotic cells.

THE ORIGIN AND DIVERSIFICATION OF MOBILE, SENTIENT MANY-CELLED ANIMALS

The closest one-celled relatives of metazoans look like the collar cells whose flagella drive water through a sponge's canals. Sponges are the most primitive grade of metazoans, yet they do things no single cell can (plate 5.2). Most suck large volumes of water through an intricate system of canals, flagellar chambers (about 10,000 of them per cubic millimeter of sponge wall), and oscula, absorbing oxygen from the water, and shedding carbon dioxide and digestive wastes into it. In the flagellar chambers, collar cells snare microbes from the passing water. Sponges, however, all lack muscles, nerves, digestive systems, and blood vessels. Adult sponges cannot move. When sponges began oxygenating the oceans by turning cyanobacteria into turds that were often buried unoxidized, they cleared the water, allowing larger, more light-demanding eukaryotic plankton (one-celled algae)

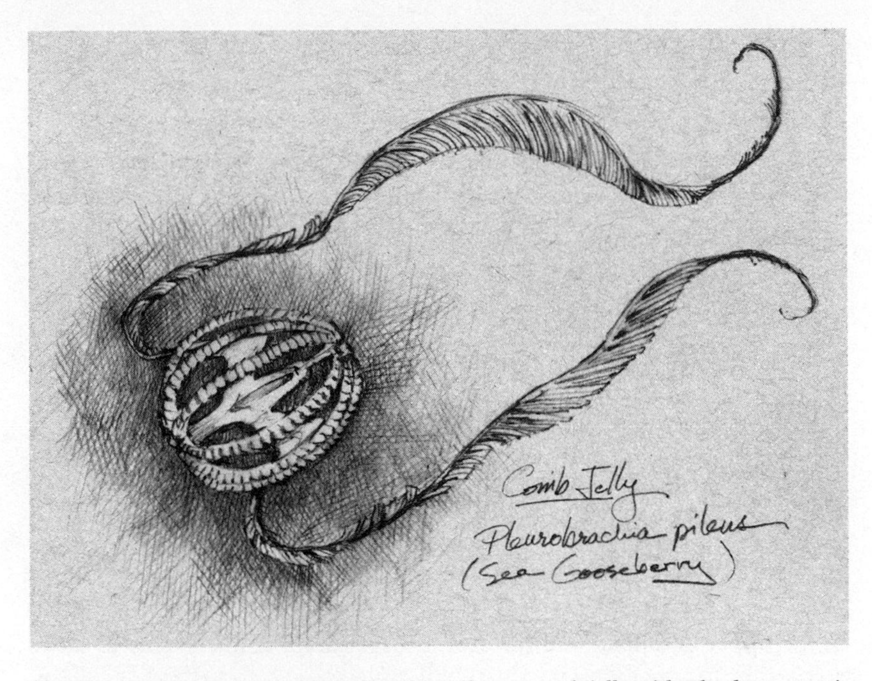

Fig. 5.2. A sea gooseberry, *Pleurobrachia pileus,* a comb jelly with a body 2.5 centimeters long and tentacles of 40 centimeters. They eat actively swimming zooplankton. Unlike sponges, ctenophores have nerves and muscles, and can choose what direction they move. Some experts think they were the earliest surviving group of metazoans to diverge. (Drawing by Debby Cotter Kaspari)

to replace cyanobacteria. This plankton provided food that other animals evolved to eat.

Among these animals were comb jellies, ctenophores: predators, often ellipsoidal, with a mouth, digestive cavity, muscles, nerves, and two long tentacles, some of whose cells secrete glue when they touch prey (fig. 5.2). They are propelled by eight rows of cilia extending from their mouths to their hinder poles. Comb jellies can choose where they move, catch prey, and migrate between their daytime and nighttime habitat. Although modern comb jellies are carnivorous feeders on algae-eaters, early ones must have eaten eukaryotic plankton, turning them into turds large enough to reach the sea bottom undecayed. By doing so, they helped to further the oceans' oxygenation.

Other simple animals with muscles and nerve nets are cnidarians—hydras, sea anemones (plate 5.3), corals, and jellyfish (plate 5.4). A hydra is a basic cnidarian unit, a "polyp": a digestive cavity, attached to the bottom, rimmed by tentacles with stinging "thread-cells." When touched, thread cells eject barbed, poison-filled threads, paralyzing prey; their tentacles draw it into the cavity, where enzymes digest it. A jellyfish is an inverted polyp, or "medusa," a pulsating bell from whose rim dangle long tentacles, full of thread cells. Its nerve net and muscle cells (which also serve as digestive cells) coordinate the bell's pulsations to keep the jellyfish from sinking. A sea anemone is a larger, stouter polyp. A coral is a colony of sea anemones, each sunk in a protective limestone cup that it secretes: they exchange nourishment by strands of tissue. Colonial corals can form huge reefs. Australia's Great Barrier Reef is visible from outer space.

Cnidarians also helped oxygenate the ocean by turning microbes into fecal pellets that fell undecayed to the sea floor. The increased oxygen in the upper ocean allowed the evolution of larger animals, able to move faster, procure more elusive food, deal with predators, choose appropriate mates, and extend their habitat by oxygenating even more of the ocean. After sponges, ctenophores and cnidarians diverged, metazoans gave rise to Bilateria, bilaterally symmetric animals with a definite front and rear, whose two sides were normally mirror images of each other. Tracks like those of burrowing bilaterian worms were fossilized 560 million years ago. Judging by how their brains and nerve cords develop in their embryos, and homologies among the genes involved, common ancestors of modern squid, bees, and people had nervous systems with sensory nerves for transmitting stimuli from outside, motor nerves for stimulating muscles to contract, primitive brains that coordinated appropriate responses to sensory inputs, and genes programming light-sensitive cells. In many groups, these evolved independently into eyes. Other genes promoted differentiation along the animal's long axis. Regulator genes, much the same for all bilateria, jointly governed where each promotor gene acted. These ancestors moved, and responded appropriately to diverse environmental stimuli; they had the beginnings of awareness and responsiveness. Before the Cambrian began, 542 million years ago, the Bilateria had split into three major divisions, each

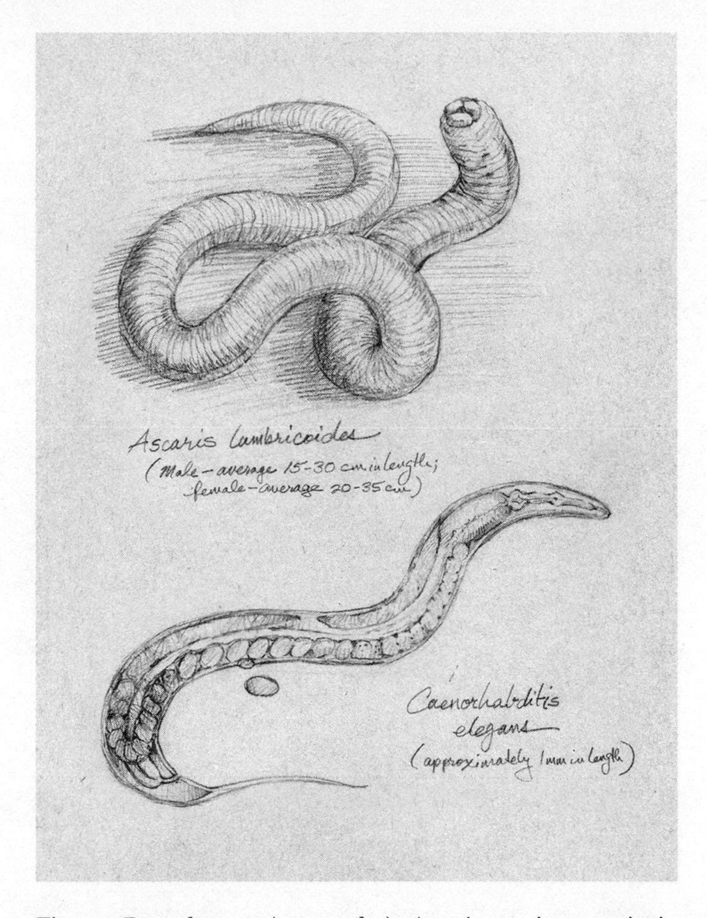

Fig. 5.3. Roundworms (nematodes). *Ascaris,* top, is a parasite in human guts; females can grow to nearly one meter long. *Caenorhabditis elegans,* bottom, is a soil nematode beloved by developmental biologists, for it has only a thousand cells and matures in three days, yet it has a nervous system and complex division of labor among its cells. (Drawing by Debby Cotter Kaspari)

with small, simple animals, which all evolved large descendants with agile, effective responses.

The division Ecdysozoa includes animals that molt as they grow—shedding their skin or shell, like lobsters, to grow larger ones. They range from nematodes (fig. 5.3), roundworms such as the gut parasite *Ascaris*

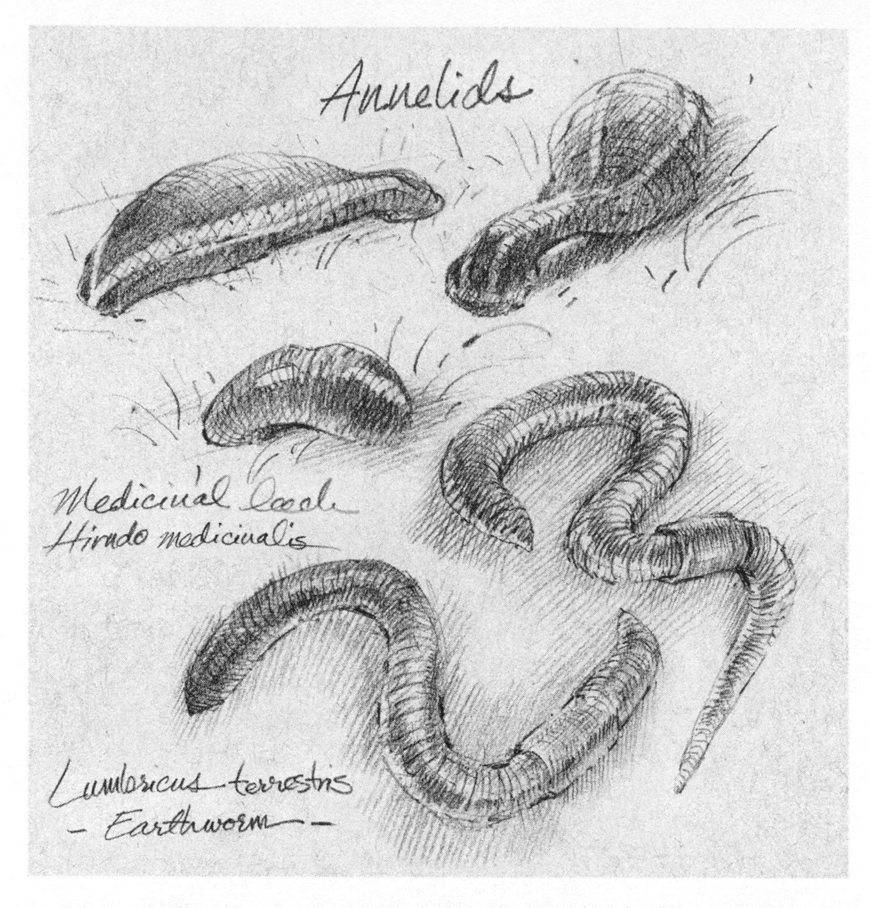

Fig. 5.4. Similar technology, different uses: Leeches (top), distant relatives of earthworms, that swim through water and suck blood from larger animals; and earthworms (bottom). Both are classified as annelids. (Drawing by Debby Cotter Kaspari)

and the 1-millimeter soil nematode *Caenorhabditis elegans,* to segmented arthropods, including agile flies. The segmentation of arthropods (and annelids, fig. 5.4, such as earthworms and leeches, and ancestral vertebrates) into repeated units allowed division of labor among segments and relatively independent modification of each segment, as multicellularity allowed division of labor and differentiation among cells. In many arthropods, individual segments and their appendages acquired different functions (plates

5.5, 5.6). The first pair of appendages became antennae for sensing the environment, the next few pairs, jaws and mouthparts; behind were legs for walking. Arthropods now range from shrimp and mantids to spiders that build prey-catching webs of elasticity and strength no equally lightweight structure can match, and dragonflies that have large compound eyes and catch flying insects by extrapolating their paths (plates 5.7–5.9). Honeybee societies are famous for the automatic, flexible, intricately coordinated division of labor among their workers.

Lophotrochozoa, many sharing similar planktonic larvae, now range from flatworms like *Planaria,* kept small and flat by their lack of means for circulating blood and oxygen, to annelids like earthworms, and mollusks, most with shells that grow at one margin as the animal grows (plates 5.10–5.13). Mollusks range from grazing chitons, snails, and immobile, filter-feeding mussels to squid and octopi. These last lack shells, but have single-chamber eyes like ours (not compound eyes like dragonflies), circulatory systems where a network of capillaries connect arteries with veins, and large, complex nervous systems—features only vertebrates match. Squid and octopi have eight or ten prehensile tentacles, capable of delicate manipulation, surrounding their heads. Octopi are notoriously clever. Squid and octopi control their color patterns. They can match their background precisely. Squid deploy a huge variety of color patterns: one side might signal courtship to a nearby female, while the other disguises itself from a predator.

Finally, deuterostomes now range from acoels, flatworm lookalikes, to starfish, sea urchins, sac-like filter-feeding tunicates, with tadpole-like larvae and clamlike intake and exhalant siphons, and vertebrates ranging from fish and frogs to birds and ourselves.

Many organisms without shells, the "Ediacaran biota," appeared 580 million years ago. Some formed reefs. Just how they lived, or even whether they were all marine, is far from clear. Few of them could move. We know none that had eyes. The only ones experts think they can recognize are: a sponge; a solitary ancestor of colonial bryozoan lophotrochozoans with a lophophore, a ring of ciliated tentacles generating currents to bring in food; a shell-less mollusk, *Kimberella,* that grazed with limpet-like teeth on algal mats covering the sea floor; and *Dickinsonia,* an animal that apparently crawled on its digestive surface, like modern *Trichoplax. Trichoplax* is very simple, with six kinds of cells (some sponge larvae have twelve kinds, while a human being has over a hundred kinds), but it coordinates these cells well enough to make an effective living. *Trichoplax* glides over the bottom by the beating of cilia on its underside. When it reaches a patch of algae, its cilia stop beating, it stops, and those of its "lipophil cells" that touch algae release a chemical that breaks algal cell walls, releasing their contents, which the animal digests. Nearly all of the Ediacaran biota disappeared just before the Cambrian, when effective predators appeared. To find and catch prey, predators evolved ever greater mobility, ever greater awareness of their surroundings, and ever more effective responses. Moreover, prey with similar abilities were among the best at escaping these predators.

During the Cambrian's 50 million years, modern ecosystems evolved. They were organized by predators that limited the abundance and distribution of their prey and recycled resources more quickly, enhancing their ecosystem's production of living matter. They formed a new world of "top-down" control and rapid evolution, very unlike the glacially slow evolution of cyanobacteria-dominated ecosystems that lasted until less than 800 million years ago. When the Cambrian began, 540 million years ago, shells, presumably anti-predator armor, of modern-day groups suddenly became abundant. At its end, all major animal groups had appeared—early arthropods, such as crawling, deposit-feeding trilobites and predaceous swimmers; snails, clams, and minute, shelled ancestors of squid; stalked crinoids related to sea urchins, and swimming vertebrates 3 centimeters long with backbones, single-chamber eyes, and brains. Burrowers abounded, recycling buried carbon, which in turn enhanced production of living matter. Many of these animals were immobile, such as armored reef-forming sponges (archaeocyathids), early corals, and early crinoids, whose stalks supported several arms that snared small organisms in the water and carried them to the mouth atop the stalk. Others moved: some crawled, others swam. Many had eyes. The swimming meter-long predator *Anomalocaris* (fig. 5.5) had compound eyes 3 centimeters wide with 16,000 facets apiece (a modern dragonfly eye has 28,000). Such compound eyes offer far less pre-

Fig. 5.5. A Cambrian scene: an *Anomalocaris* swimming near the sea bottom, examining potential prey. Its good eyes allowed it to view its surroundings. It could see, recognize, seize, and eat mobile prey. (Drawing by Debby Cotter Kaspari)

cise images than our single-chamber eyes, but they offer panoramic vision, and detect moving objects such as prey or predators more readily. Vision, brain, optic nerves, and nerves of the front appendages coevolved as successive animals became better able to sense and interpret their surroundings, and coordinate appropriate responses. Indeed, designing "intelligent robots" with sensors, motor organs, and a control system that coordinates appropriate motor responses to sensory data have helped us understand the sensory-motor coordination an insect needs to fly through cluttered environments, and how ancestral salamanders might modify their forebears' swimming behavior in order to walk. In sum, predation prompted the evolution of animals increasingly more aware of, and better able to cope with, their surroundings. The diversity of marine life kept exploding for 40 million years beyond the Cambrian.

TRADE-OFFS AND DIVERSIFICATION

These innovations sparked a planet-wide diversification of organisms big enough for people to see. Living things diversified because no one kind of life can do all things well; as for people, "the jack of all trades is master of none." To be specific, organisms face trade-offs, whereby enhancing one ability necessarily diminishes another.

There is a trade-off between the ability to catch prey of very different sizes; jaguars that normally live on deer or peccaries would be hard put to live on mice, whereas house-cat-sized margays that can live on mice cannot handle deer. Therefore most communities have several different sizes of each type of predator.

Likewise, the ability to defend rich resources trades off against the ability to live on scarce ones. Animals that defend rich food sources from competitors must eat enough to maintain this ability; they cannot survive where food is scarce, whereas animals that live on scarce resources cannot maintain the vigor and quickness needed to defend rich sources. The biologist Ross Robertson demonstrated this trade-off within a guild of three territorial damselfish common on Caribbean coral reefs. When he removed the dominant species from a reef, the two smaller species doubled in numbers over the next four years, thanks to eating food the dominants formerly defended. The dominants, however, could not live on the meager resources released by removing the smaller species, so removing them did not lead to increase in the dominants' numbers.

Animals and plants all face trade-offs between fast growth and effective defense. Sponges abound on both coral reefs and the arching roots of red mangroves, *Rhizophora,* the trees that fringe many tropical shores.

These roots are flooded at high tide. Species of sponges common on coral reefs, however, rarely grow on mangrove roots, and vice versa. Coral reefs offer sponge-eating fishes many places to hide, so only poisonous sponges survive there. Mangroves offer sponge-eaters far fewer hide-outs, so mangrove sponges grow faster because they have less need to invest in defensive toxins. Sponge-eating fishes eat tastier mangrove sponges if they are transplanted to reefs, whereas faster-growing mangrove sponges overgrow and smother reef sponges transplanted to mangroves.

Trade-Offs and Community Organization

Such trade-offs favor a diversity of life that exploits available energy ever more effectively and supports ever more kinds of microbes, plants, and animals. They also drive divergence among ecosystems. The oldest diverse ecosystems we know from fossils are marine. How can fossils reveal what trade-offs drove divergence among past ecosystems?

NEAR-SHORE MARINE COMMUNITIES: AN ORGANIZING TRADE-OFF

To learn what factors organize fossil communities, we must learn how trade-offs organize their modern counterparts. One trade-off shaping near-shore marine communities is that of fast growth where resources are abundant versus effective defense against predators or herbivores where resources are scarce. This trade-off, which is related to the one driving the contrast between the well-defended, slow-growing sponges on coral reefs and the faster-growing, more edible sponges on mangrove roots, organizes wave-beaten communities of rocky shores in the northeastern Pacific. Robert Paine, a professor of zoology at the University of Washington, devoted his life to studying communities of rocky shores at Tatoosh, a group of islets bearing the lighthouse for Cape Flattery, the northwest tip of the Olympic Peninsula in Washington state. Life on these shores forms distinct zones. Unlike the land, where primary space-holders are always plants, a plant zone often borders a zone of immobile filter-feeding animals (plates 5.14,

5.15). Highest is a green wash of microscopic algae occasionally splashed by the waves below. Below them is a zone of barnacles, filter-feeders, sometimes submerged by surges of blue water. A zone of inch-high upright algal blades, *Mazzaella,* forming a springy golden-green carpet, often separates the "barnacle zone" from a zone of dense mussel beds. Broad-bladed kelps occupy the lowest part of the "intertidal," which is sometimes above, sometimes under water, as the tide ebbs and floods. The kelps are protected by starfish that eat mussels that would otherwise spread down and replace them (plate 5.16).

Grazing sea urchins aggregate in pools where waves cannot dislodge them (plate 5.17). Only slow-growing crustose coralline algae with photosynthetic cells embedded in their limestone crust can survive these urchins' grazing; urchins keep out faster-growing kelps. In contrast, kelps grow fastest where waves beat hardest, allowing sea urchins and other grazers little time to move and eat without being knocked away. On exposed angles, waves tear great, long-lasting gaps in the mussel bed where sea palms, *Postelsia palmaeformis,* kelps with a thatch of narrow fronds atop a springy, hollow stalk, sprout up each spring, forming dense stands up to 60 centimeters tall (plate 5.18). These stands grow so fast because their water is rich in nutrients and waves protect them from herbivores, so they need not make defensive chemicals. Waves also yank out sea palms that grow tall enough to shade their neighbors. Waves stir their lightweight fronds, dividing light more evenly among them than light is ever divided among a tall forest's leaves, so *Postelsia* stands can support more than 15 square meters of fronds per square meter of rock, compared to a rain forest's 6–8 square meters of leaves per square meter of ground. During their growing season, this extra frond area allows these *Postelsia* to produce vegetable matter three times faster than a rain forest. By November, however, storms rip these kelps away.

TRADE-OFFS AND MARINE COMMUNITIES ON PANAMA'S TWO COASTS

The trade-off between fast growth and effective defense also drives the contrast between Panama's Caribbean shore, with its clear water and abun-

Fig. 5.6. A massive Caribbean coral reef in Bocas del Toro, western Panama. Such coral reefs thrive in clear, unproductive water. (Photograph by Christian Ziegler)

dant reefs, and the far more nutrient-rich settings in the Pacific-side Bay of Panama. On the Caribbean side, the clear water is poor in nutrients, corals thrive, and the sea bottom is sand, not mud. Only where rivers spill floods of muddy nutrient-rich water into the sea are things different. Nevertheless, even away from river mouths, Caribbean shores are heterogeneous. In many places, offshore reefs front the open ocean, breaking the force of its waves (fig. 5.6). The calm water behind harbors seagrass beds and sand-bottomed lagoons spotted with patch reefs, and a shore with mangroves, trees tolerant of tidal flooding by salt water, which shelter young fish whose adults keep corals free of algae, and trap mud from terrestrial runoff that would otherwise smother corals and seagrass beds alike. In this array of interdependent communities, nutrients are scarcest over the reef and most abundant near the mangroves. Elsewhere, a reef flat extends from shore to a reef crest fronting the ocean. Here, we focus on coral reefs, the extreme in effective defense.

On Caribbean shores receiving the full force of wind and wave, massive coral reefs often front the sea. Crustose coralline algae, with well-protected photosynthetic cells, form the wave-beaten reef crest; elsewhere,

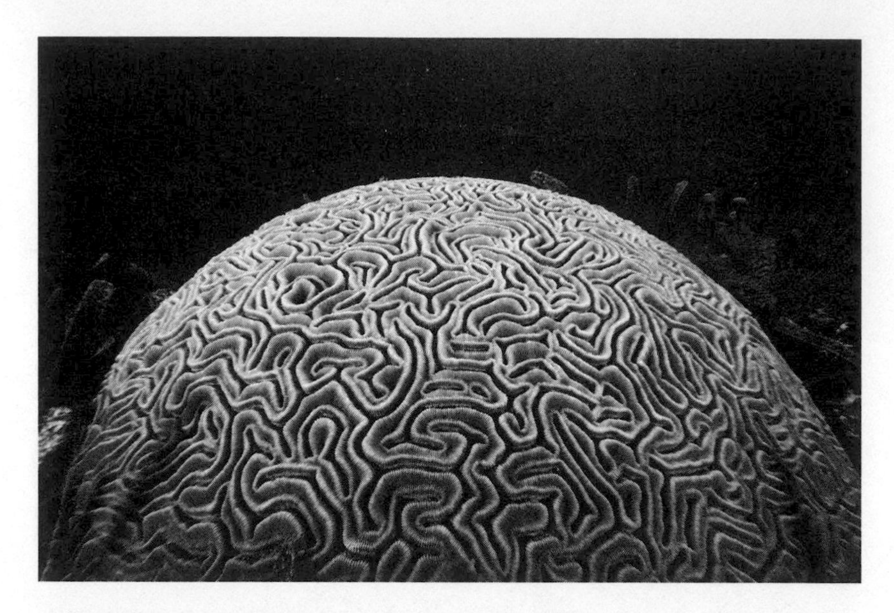

Fig. 5.7. The trade-off between fast growth and effective defense. On the reef in Panama's San Blas Islands, algal cells live in coral polyps sunk in protective limestone cups on the surface of colonial corals such as the *Diploria labyrinthiformis* shown here. Unproductive Caribbean waters favor effective defense. (Photograph by Christian Ziegler)

they help cement corals together into a cohesive, wave-resistant reef framework. Because nutrients such as nitrogen and phosphorus are scarce in sunlit surface waters, phytoplankton—microscopic algae—are far too sparse to compromise the water's clarity. Newly settled corals prosper because neither sediment nor fast-growing algae smother them. These corals cannot live only on the animals their polyps can snare from the water; so they harbor algae in their polyps that supply carbohydrates in return for nutrients released from animals the polyps digest, and a well-lit home safer from algae-eating fish (fig. 5.7). These corals form reefs full of crevices and nooks that hide eels, worms, and plant-eating, sponge-eating, and some coral-crunching fishes from their predators. Slow-growing, barely edible sponges filter microbes from the water, keeping it crystal clear except when waves stir up the sterile sands. Farther from shore, and the fertilizing nutrients draining from it, the water harbors fewer microbes. Here, some sponges have symbiotic bacteria that extract dissolved organic matter from the water, which the sponges' flagellated collar cells cannot do. On Australia's Great Barrier Reef, nutrients are even scarcer in open water than in the Caribbean, and the sponges farthest offshore depend on symbiotic cyanobacteria for their carbohydrates. Animals of coral reefs live long, grow slowly, and have few young: coral reefs do not house sustainable commercial fisheries.

Panama's Pacific shore is much richer in nutrients. Every dry season, trade winds blow from the Caribbean across the low-lying isthmus to the Bay of Panama. These winds blow the surface water of the Pacific offshore: cooler, nutrient-rich water welling up from the sea bottom replaces it. These upwellings fertilize a rich growth of phytoplankton, microscopic algae, at the sea surface. Zooplankton such as copepods eat the algae; fish such as anchovies eat the zooplankton; great flocks of pelicans and boobies eat enough anchovies to raise abundant young. This planktonic bounty depends on the trade winds, so it is seasonal. Smaller fish are accordingly short-lived, and larger fish cannot afford to specialize only on a few prey. The prolific sea surface creates a steady rain of organic matter: dead plankton, feces of bird and fish, and so on.

At most sites in the Bay of Panama, corals cannot recruit. In shallow water with enough light to support zooxanthellae, fast-growing algal filaments or the sediment they trap smother corals lucky enough to find bare rock to settle on. Here, reefs are scattered and small. With no barrier reefs to protect them, seagrass beds are absent, and mangroves are confined to the wide mouths of large rivers, where ocean waves do not reach.

Instead of reefs, offshore mudflats—a paradise for shell collectors—support a rich biota of clams, worms, snails, and other animals that filter plankton and their remains from the seawater or burrow through the mud, eating organic detritus or smaller animals living on it. Larger snails eat smaller snails, clams, and worms; stingrays work the bottom for edible prey. Unlike those of coral reefs, animals in this world of seasonal abundance are short-lived, grow fast, and reproduce quickly and copiously. Upwellings, which occupy less than 0.01 percent of the ocean's surface, produce half the world's catch of sea fish.

HOW TO DETECT AN ORGANIZING TRADE-OFF FROM THE FOSSIL RECORD

Fossils from Panama also show how nutrient availability drives community organization. When the isthmus of Panama joined the Americas three million years ago, it snuffed out the upwellings on Panama's Caribbean side, and coral reefs and calcareous algae replaced assemblages of clams and snails. How do we know this?

The answer comes from cupuladriid bryozoans, well-integrated colonies whose component individuals, minute filter-feeders, are called zooids. Cupuladriids live on the sea floor: their colonies are shaped like shallow coolie hats, perhaps 13 millimeters across. In a cupuladriid colony, each filter-feeding "zooid," set in its own calcareous box with surface area about 0.1 square millimeter, is paired with a non-feeding zooid with a mobile hair or seta over a millimeter long. If buried, cupuladriids excavate themselves with their setae.

Feeding zooids grow larger in colder water. Colonies experiencing large seasonal changes in temperature will have both large and small zooids. Thus variation in size of feeding zooids can be used to measure the seasonal range of temperature experienced by the bryozoan colony. In Panama, the seasonal range of sea temperature is greater where seasonal cold-water upwellings are more vigorous. Thus a colony's zooids vary more in size where upwellings are stronger and nutrients more abundant (fig. 5.8).

When he was a postdoctoral fellow at the Smithsonian Tropical Research Institute in Panama, Aaron O'Dea used variation in sizes of feeding zooids in fossil cupuladriid colonies to measure the strength of upwelling. First he assessed variation in sizes of feeding zooids within cupuladriid colonies in samples dredged from the sea bottom on both sides of Panama. Colonies from the Caribbean side, where there are no upwellings and sea temperature changes little, had feeding zooids of nearly uniform size. Samples containing these bryozoans were full of fragments of coral and calcareous algae, and had few clams or snails. In colonies from the Bay of Panama, seasonal cold-water nutrient-rich upwellings create variation in sizes of feeding zooids. Pacific samples were full of clams and snails, but had few fragments of coral or calcareous algae.

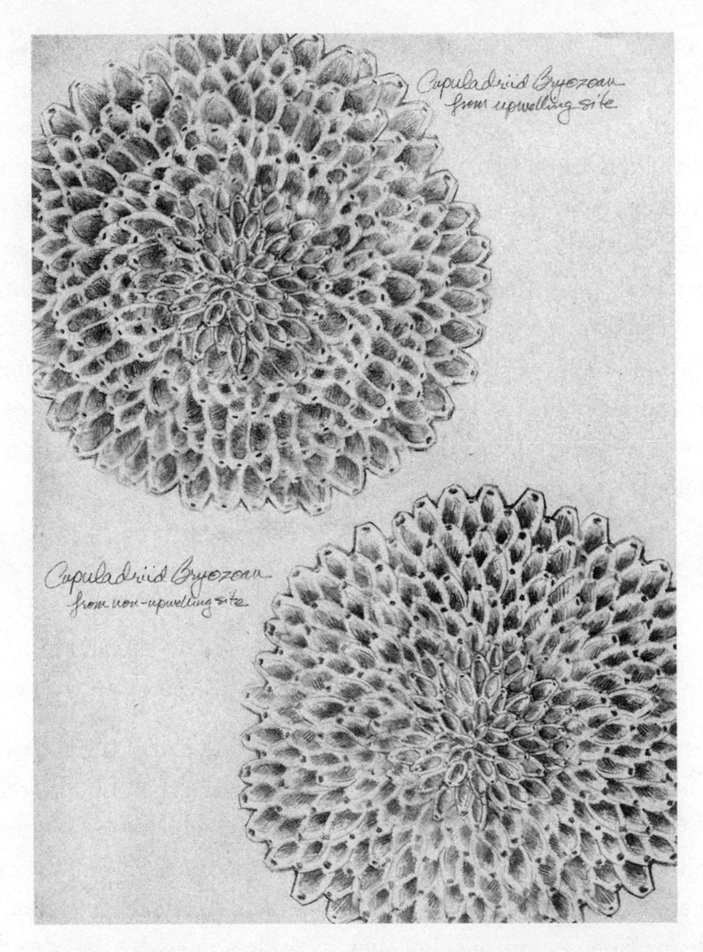

Fig. 5.8. The present as a key to the past: Cupuladriid bryozoans, a colony of which forms a wide conical cap about a half inch (1.3 centimeters) across, made up of many "cells." They make larger "cells" for their zooids when their surrounding water is cooler. Closeups of cupuladriids from upwelling and non-upwelling sites: those experiencing seasonal upwellings of cool, nutrient-rich water have zooids of more varied size. (Drawing by Debby Cotter Kaspari)

O'Dea then took samples from fossil beds up to 10 million years old along Panama's Caribbean coast. Variation of feeding zooid size in cupuladriid colonies over 4.5 million years old indicated that vigorous seasonal upwellings prevailed then. The samples from which they were collected were full of clams and snails, with few fragments of coral and calcareous algae. The decline in variability of feeding zooid size within bryozoan colonies suggests that upwellings ceased in the Caribbean three million years ago, when the isthmus finally connected the two continents. Later samples suggest that corals and calcareous algae had largely replaced clams and snails. Thus, snuffing out Caribbean upwellings caused communities organized for deterring predators and coping with nutrient scarcity to replace communities organized for fast growth.

The Evolution of Land Plants

TRADE-OFFS BETWEEN LIVING IN WATER AND LIVING ON LAND

Now let us turn to what allowed plants to colonize the land. For about two billion years after bacteria evolved photosynthesis, the only life on land was soil microbes, some of which were photosynthetic bacteria. There was enough oxygen in the air 2.3 billion years ago to form an ozone layer in the upper atmosphere, blocking ultraviolet light. This made the land more hospitable for larger organisms, but they were slow to evolve, because larger plants face major trade-offs between sea and land life:

- The kinds of chlorophyll best suited to catch photons underwater are less effective in the open air.
- Water plants are buoyed up by the water they live in, whereas land plants must support themselves. Thus seaweeds have flexible cell walls that allow them to bend, not break, under the waves, whereas land plants need stiff cell walls to stand upright.
- The surfaces of water plants allow free entry of nutrients and carbon dioxide and free exit of wastes, whereas land plants

must separate themselves from the air by a watertight cuticle—a transparent waxy "skin"—to keep their cells' watery contents from drying out. They need structures through which to discard wastes, acquire carbon dioxide, and bring water and nutrients to their cells from the soil or elsewhere.

- Sexual reproduction is much more difficult for land plants. In sexual reproduction, haploid gametes (often produced by many-celled haploid gametophytes), each with a single set of genes, unite, as do an egg and sperm, to form a diploid zygote with two sets of genes, which may divide again and again to form a many-celled sporophyte, which in mosses parasitizes the photosynthetic gametophyte. Eventually, some "sex cells" divide, producing haploid cells with one set of genes apiece. In some plants, these cells grow to be many-celled haploid gametophytes, so generations of gametophytes and sporophytes alternate. Gametophytes form gametes, male and female, which join to make diploid zygotes that become the next generation's sporophytes. In aquatic algae, "sperm" could swim to the eggs. Circumventing this need to swim is a major theme in the saga of the evolution of land plants.
- Finally, spores of water plants stay wet, but spores of the earliest land plants, like those of mosses and ferns today, had to be small and watertight enough to drift like dust in the air without their contents drying out.

HOW PLANTS ADJUSTED TO LIVING ON LAND

We now confront two questions: First, who were the ancestors of land plants? Green algae such as sea lettuce, *Ulva,* and the filaments of *Spirogyra* found in fresh water, not red algae like *Mazzaella* or brown algae like the kelps of the weather coast of Tatoosh, are their ancestral group (plates 5.19–5.21). Like land plants, green algae store starch in their chloroplasts, and have chlorophyll b, the chlorophyll best suited for open-air light. Many features identify charophytes, a group of freshwater green algae, whose only many-celled stage is gametophytes, as ancestors of all land plants. They have stiff cell walls; some stand upright. The charophyte *Coleochaete* coats its zygotes with a waterproofing agent, sporopollenin, to prevent their drying out when exposed to air, which land plants also use to protect their spores and pollen. *Colochaete* keep their eggs in minute pockets, archegonia, on their gametophytes' surface, where swimming sperm fertilize them. E. J. H. Corner remarked that "If two persons wish to find each other it is better for one to wait while the other searches. And if on meeting they must journey, it is better if she who waits should be provisioned while he that searches may travel light and fast. The principle of assignation was worked out long ago by gametes. . . . It is the way of all higher green plants of the land and of the fresh water green [algae]." Some *Coleochaete,* moreover, retain and nourish the fertilized egg in its archegonial pocket as mosses do, giving it an advantage over spores lacking such help. The first land plants apparently evolved from relatives of *Coleochaete.*

Second, how did plants adapt to life on land? When plants first colonized the land, fungi already present formed symbioses with them, providing plant hosts with nutrients drawn from the soil by their thread-like hyphae in return for carbohydrates. Liverworts (plate 5.22) first diverged from ancestors of other land plants (that is to say, liverworts are the "sister group" of all other land plants). Attracting fungi that supplied them with nutrients in return for carbohydrates, as mycorrhizae do today, almost certainly enabled early liverworts to invade the land. Judging by the abundance of tetrahedral groups of four spores apiece, like those of today's primitive liverwort *Sphaerocarpos,* liverworts dispersed by spores. They spread over the land by the mid-Ordovician, 465 million years ago, locking up enough carbon dioxide to cause severe glaciation. Mosses diverged next (plate 5.23). No modern moss employs fungi to extract nutrients from the soil, perhaps because early mosses, like modern *Sphagnum,* lived in nutrient-poor sites and extracted nutrients from rainwater. Mosses are the sister group to hornworts plus all vascular plants (club mosses, ferns, horsetails, and seed plants). Like liverworts, most hornworts and vascular plants need mycorrhiza-like fungi to extract nutrients from the soil.

As we shall see, coping with land life shaped plant evolution for 300 million years. The water content of many modern liverworts and mosses, like that of the earliest land plants, reflects the wetness of their environment, just as a lizard's temperature reflects its environment's temperature. Even if thoroughly dried out, these plants recover normal function when rehydrated. In contrast, cells of most land plants (and all animals) live in a wet setting a bit like the seawater in which their ancestors lived long ago. They are "homeohydric": their body's water content is kept constant, regardless of how dry their surroundings are. They die if they dry out. How did they become homeohydric?

About 440 million years ago, to judge by the appearance and spread of their characteristic single spores, vascular plants began replacing the liverworts and mosses that first covered the earth. Traces of protective cuticle, and of tubes less than 0.1 millimeter wide, including tracheids with reinforced walls, in which plants moved water from the soil to their shoots—signs of homeohydry—appeared 430 million years ago. The first *Cooksonia* (fig. 5.9) appeared then, fossilized only as non-photosynthetic sporophytes less than 5 centimeters (2 inches) tall, like those of mosses except that they branched dichotomously, with sporangia atop their branch tips. Later, larger, photosynthetic *Cooksonia* sporophytes had stomates, pores in their stem cuticle that let in carbon dioxide, at the cost of losing water vapor. A plant's stomates closed if it lost too much water. Then, the atmosphere harbored ten times as much carbon dioxide as today, so plants lost less water to acquire it. Roots distinct from stems appeared about 400 million years ago in *Rhynia* (fig. 5.10). Its dichotomously branched shoots, up to 3 centimeters (about an inch) thick and 18 centimeters (7 inches) tall, grew from horizontal rhizomes, underground roots with thread-like rhizoids (root hairs).

Ancestors of lycopsids, a group of vascular plants including modern club mosses, *Lycopodium,* and *Selaginella* (plate 5.24), diverged from ancestors of horsetails, ferns, and seedplants over 410 million years ago. *Zosterophyllum,* related to ancestral lycopsids, had horizontal rhizomes from which grew branched stems 1–4 millimeters thick and a foot (30 centimeters) tall, with sporangia along their upper parts. These stems had a solid core of stiff-walled tracheids, vessels for transporting water. At that time,

Fig. 5.9. A sporophyte, with spore-bearing containers at the tips of its thin branches, 7 centimeters tall, of *Cooksonia,* the first known genus of fossil vascular plant. It evolved about 430 million years ago. *Cooksonia* appears to be part of the group ancestral to all living vascular plants. Like moss sporophytes, the oldest *Cooksonia* sporophytes did not photosynthesize. (Drawing by Debby Cotter Kaspari)

leafless shrubs ancestral to ferns and seed plants, *Psilophyton,* up to 60 centimeters (2 feet) tall, likewise moved water from the soil through a core of conifer-like tracheids.

Soon after mosses and liverworts spread over the earth, insects that ate their spores, or at least insects with turds full of spores, had evolved from springtail-like ancestors, and nematodes colonized the soil. *Rhynia*

Fig. 5.10. *Rhynia,* the first land plant with differentiated roots, 20 centimeters tall; it lived over 400 million years ago. (Drawing by Debby Cotter Kaspari)

suffered from minute insect pests. Some pierced stems to suck interior fluids, others were stem-borers, yet others chewed on the stem's photosynthetic exterior. Other insects living among these plants ate detritus.

THE EVOLUTION OF TREES

A small, upright leafless stem should place its photosynthetic tissue just under its cuticle and its support and supply tissue within, as does *Zosterophyllum.* Larger plants, however, reduce the amount of support tissue they need by deploying it on the plant's outer rim, like a bamboo or palm. Doing so favors a division of labor among leaves, the organs of photosynthesis; the trunk and branches supporting these leaves; and the roots that anchor the tree and take up the water and nutrients it needs. By 385 million years ago, trees had solved these problems of support and supply, thereby spawning a "tragedy of the commons," where competition to lift leaves above one's neighbors' made it ever harder for plants to place their leaves in full sun.

A prominent early tree was *Archaeopteris* (fig. 5.11), closely related to ancestors of modern trees. *Archaeopteris* looked very modern. It was 10 meters (30 feet) tall, with gingko-like leaves on spirally arranged branches and stout, branching roots up to 1 meter (40 inches) deep. Like modern trees, their diameter kept increasing: in seasonal climates, they formed annual rings. *Archaeopteris* dispersed spores, which formed minute gametophytes, a bit like those of ferns (fig. 5.12). As in mosses and ferns, sperm had to swim through a film of water to fertilize the egg. Once fertilized, a gametophyte's egg grew into a sporophyte tree. These trees' wood, like that of modern conifers, had tracheids 0.1 millimeter wide and 2 centimeters long, through which water moved from roots to leaves, and phloem vessels, carrying nourishment where it was needed. These trees used energy from sunlight to pull water up from its roots to its leaves. As water evaporated, or transpired, through the stomates, capillary traction pulled more water from below. Such traction places the water under tension. If air is sucked into one of these tracheids, it fills the tracheid, causing an embolism that "snaps" the pull, just as a tow ceases when the cable snaps by which the tugboat tows its barge. On a hot, sunny day one can hear the "snaps" these embolisms cause by placing a stethoscope on a tree trunk. Transpiration consumes over a third of the energy from incoming sunlight, far more than photosynthesis does. Transpiration keeps the forest cool, and spawns rain clouds that keep it wet. As E. J. H. Corner remarked, "transpiration is the overwhelming activity of the forest, which through plenteous evaporation engenders its own storms."

Archaeopteris had bizarre company. By 375 million years ago, some lycopsids, whose ancestors had diverged from those of ferns and seed plants 35 million years earlier, were tall trees, soft-stemmed like giant herbs, with impressive roots. Tree ferns, and tree horsetails up to 18 meters tall, appeared by 354 million years ago.

The next step in adapting to land, seed-making, opened a new, striking chapter in plant life. Unlike spores, seeds give their plant's young a stock of resources that go with them when they disperse. Sporophytes took

Fig. 5.11. The oldest known tree, *Archaeopteris,* which lived 375 million years ago, had wood like a modern conifer's, and reproduced, like ferns, by spores. (Drawing by Debby Cotter Kaspari)

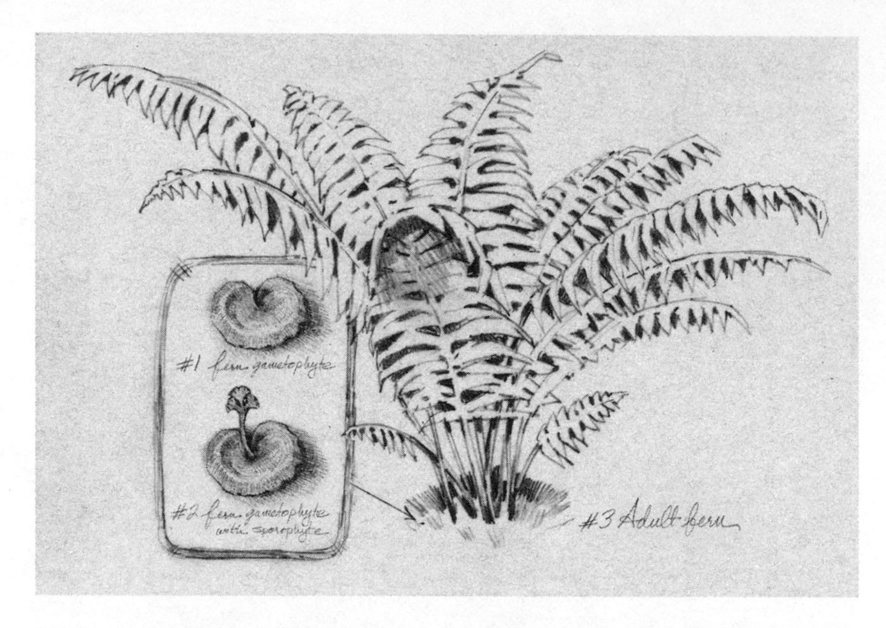

Fig. 5.12. Alternation of generations in a fern between a large sporophyte and a minute haploid gametophyte. A new sporophyte begins its life when an egg in this gametophyte is fertilized by a "sperm" from another. (Drawing by Debby Cotter Kaspari)

the first step toward seed-making by making two sizes of spore—larger megaspores that became female, egg-making gametophytes, and smaller microspores that became male, sperm-making gametophytes. Seed plant ancestors like *Archaeopteris,* club mosses, and tree horsetails all evolved this feature independently. Next evolved sporangia specialized to produce only megaspores or only microspores, megasporangia maturing only one megaspore apiece, a "seed-coat" covering each single-spore megasporangium, and finally, keeping the megasporangium on the plant until its spore, or "ovule," is fertilized, at which point it is a seed.

Seeds became especially useful when sperm could fertilize the ovules without swimming to them. First, microspores formed gametophytes that could be blown about like pollen grains, while megasporangia formed sticky "pollination drops" at their openings that adhered to and retained only male gametophytes of their species. Early on, male gametophytes landing on the right pollination drop released sperm that swam through the megasporangium to fertilize the ovule. Male gametophytes of later species, called "prepollen," formed tubes that extended into the ovule, through which the sperm swam to it: cycads and ginkgos still do this. Yet later species produced genuine pollen, which upon landing on the right pollination drop, produced "pollen tubes" delivering non-motile sperm to the ovule, as *Gnetum,* conifers, and flowering plants do to this day.

Elkinsia, the first seed-bearing gymnosperm trees with genuine pollen, appeared 360 million years ago. Seed-ferns soon followed. These plants, tree club mosses, *Lepidodendron,* and tree horsetails, *Calamites,* formed the great "coal swamps" of the Carboniferous, 350–290 million years ago (figs. 5.13, 5.14). One such swamp-forest, preserved 300 million years ago in a coal seam in southern Illinois, harbored tree ferns, *Psaronius,* and seed-ferns, *Medullosa* (figs. 5.15, 5.16). It also had tree horsetails, tree club mosses, and other trees and lianas. By this time leaf-eating insects had evolved, along with amphibians, that laid their eggs in the water where their tadpoles grew up, and early reptiles that laid eggs on land, with watertight shells to keep them from drying out. Like plants, vertebrates evolved successive stages of emancipation from life in water.

When coal swamps flourished, no organisms could break down dead wood as termites and white rot fungi do now, so dead vegetable matter fossilized. The resulting coal is now being burned to generate much of our electricity, and the power we use for our factories. The carbon locked up in this dead wood lowered the atmosphere's carbon dioxide content, cooling the world. Leaves now needed more stomates to obtain sufficient carbon dioxide from the atmosphere for photosynthesis, and their stomatal index, the proportion of these leaves' undersides covered by stomates, increased. Indeed, stomatal index of fossil leaves is now used to estimate their atmosphere's carbon dioxide content. Stomatal indices show that the atmosphere's carbon dioxide content began to decline 375 million years ago, soon after the first forests evolved, and to plummet 350 million years ago. By 300 million years ago, atmospheric carbon dioxide fell to today's levels, and glaciers spread from the poles. Stomatal index suddenly dropped 200 million years ago, at the end of the Triassic, when the atmosphere's carbon

Fig. 5.13. A giant herb, *Lepidodendron,* a tree club moss up to 30 meters tall and 1 meter in diameter, with elaborate roots, that made two kinds of spores. It lived in peat swamps 300 million years ago. These swamps are the primary source of our coal, so the time of these "coal swamps" is called the Carboniferous (coal-bearing) era. (Drawing by Debby Cotter Kaspari)

dioxide content tripled, causing sudden, severe global warming that triggered a wave of extinctions.

Although seed-bearing gymnosperms appeared 360 million years ago, and seed-ferns soon after, their heyday came after the coal-swamp era. The most widespread arboreal seed-ferns, *Glossopteris* and its relatives, dominated the southern hemisphere from 290 to 250 million years ago. Cycads

Fig. 5.14. Another giant herb, *Calamites,* a tree horsetail 15 meters tall, that lived in coal swamps 300 million years ago. (Drawing by Debby Cotter Kaspari)

and gingkos very like their modern relatives appeared 290 million years ago. Cycads were abundant all through the age of dinosaurs, which ended in a mass extinction 65 million years ago. Conifers appeared 315 million years ago, but only began to diversify in earnest 70 million years later. By 175 million years ago, conifers included ancestors of *Araucaria,* the genus of "Norfolk Island pines," redwoods (both evergreen *Sequoia* and deciduous *Metasequoia*), cypresses, and pines.

Many survivors of formerly dominant groups such as horsetails, club mosses, and ferns are now small plants, often in marginal, shady, or

Fig. 5.15. A swamp forest tree fern, *Psaronius*. (Drawing by Debby Cotter Kaspari)

swampy habitats. Nearly all survivors from these ancient forests—club mosses, horsetails, ferns, cycads, and conifers—are notorious for their anti-herbivore poisons. Indeed, as human beings sometimes poison ecosystems with the pesticides they use to defend their crops, the long-lived pesticides some conifers use to defend their leaves poison their soils in ways that slow the decomposition of leaves and wood and the recycling of the nutrients contained therein. Did these forests manifest a sort of "siege mentality" that slowed growth in order to avoid being eaten? E. J. H. Corner remarked that

Plate 1.1. Some animals eat others. This sequence of photographs shows a fishing bat, *Noctilio leporinus,* catching a fish it detected from echoes of its own ultrasonic calls, its sonar, in the Laboratory Cove of Barro Colorado Island. (Photograph by Christian Ziegler)

Plate 2.1. A mother sloth, *Bradypus variegatus,* specialized to live only on leaves, with its dependent offspring. Taking care of one's young is the origin of social life. (Photograph by Christian Ziegler)

Plate 2.2. A jumping ocelot, *Leopardus pardalis,* now the top predator on Barro Colorado Island. (Photograph by Christian Ziegler)

Plate 3.1. A central problem most animals face is how not to be eaten. The urgency of this need is revealed by how precisely some insects mimic things insect eaters do not eat. This leaf-winged katydid, the Malaysian "walking leaf," *Phyllium giganteum,* imitates a broken leaf, of no interest to insect eaters. (Photograph by Christian Ziegler)

Plate 3.2. Some edible insects "cheat" by mimicking poisonous models, thereby undermining the effect of the model's advertisement. In Panama, the edible *Mimoides ilus* (left) mimics the poisonous *Parides eurimedes* (right). (Photographs by Annette Aiello)

Plate 3.3. A disguise repeated too often forms a pattern that predators detect: it is better to vary the disguise, as do these little toads of the leaf litter, *Rhinella alata*. (Photograph by Christian Ziegler)

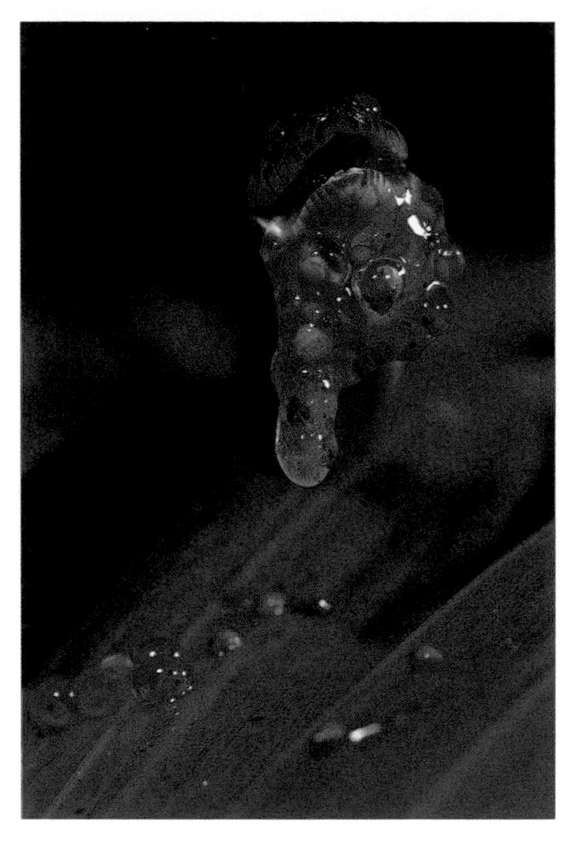

Plate 3.4. In most frogs, such as this Panamanian red-eyed tree frog, *Agalychnis callidryas,* tadpoles live in water, adults on land. (Photograph by Christian Ziegler)

Plate 3.5. Red-eyed tree frogs lay eggs on leaves overhanging water. If a snake starts eating a frog's eggs a day or two before normal hatching time, the tadpoles hatch early and drop into the water. (Photograph by Christian Ziegler)

Plate 3.6. Some animals, such as this group of long-tailed macaques, *Macaca fascicularis,* in Penang, Malaysia, find safety from predators and competitors by living in groups, which can fight them off. (Photograph by Christian Ziegler)

Plate 3.7. A colony's health depends on effective waste disposal. Workers discard used leaf fragments on the colony's dump. The toad, *Rhinella marina,* is lying in wait to eat some of these workers. (Photograph by Christian Ziegler)

Plate 5.1. A canopy of diverse tropical forest along the Panama Canal. The tree with the crown of vermilion flowers is the cuipo, *Cavanillesia platanifolia,* which flushes new leaves at the end of Panama's dry season, using water stored in its trunk. Photosynthesis provides the fuel for evolving complex communities. (Photograph by Christian Ziegler)

Plate 5.2. Coral reef sponges on a coral reef in Belize. By filtering microbes, which they eat, from the water they pump through their canals, sponges help keep their reef waters crystal clear. The yellow-orange sponge is *Mycale laevis;* the pink one is *Desmopsamma anchorata.* (Photograph by Christian Ziegler)

Plate 5.3. Stages on life's way: A clump of sea anemones, *Stichodactyla helianthus,* on mangrove roots from Panama's Caribbean coast. These are cnidarians, which have muscles and nerve nets but no brains and no anus. (Photograph by Christian Ziegler)

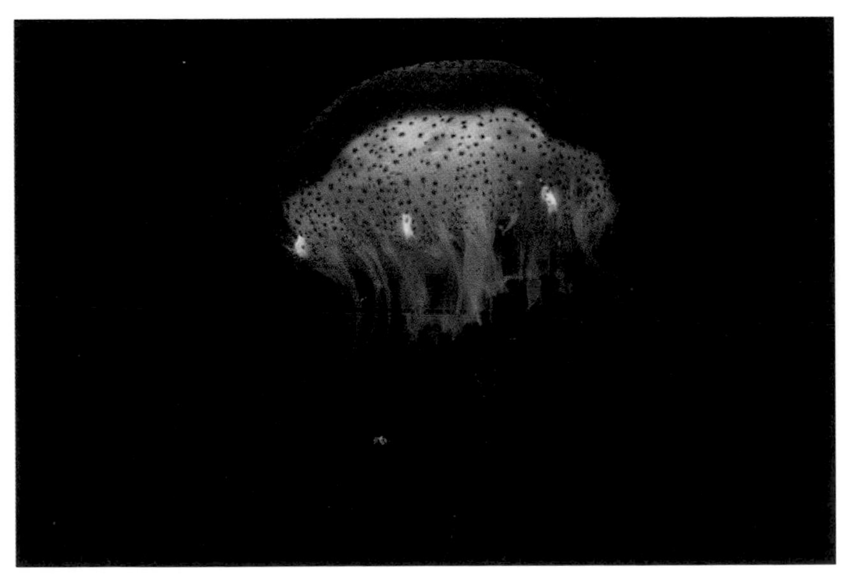

Plate 5.4. Stages on life's way: A jellyfish, *Viatrix,* another cnidarian. (Photograph by Christian Ziegler)

Plate 5.5. Stages on life's way: Segmentation in a centipede, crab, scorpion, grasshopper, and spider, with abdominal segments colored blue, thoracic segments blue-green, and cephalic (head) segments green. Segmentation enables division of labor among segments. (Drawing © 2016, Danielle Van Brabant)

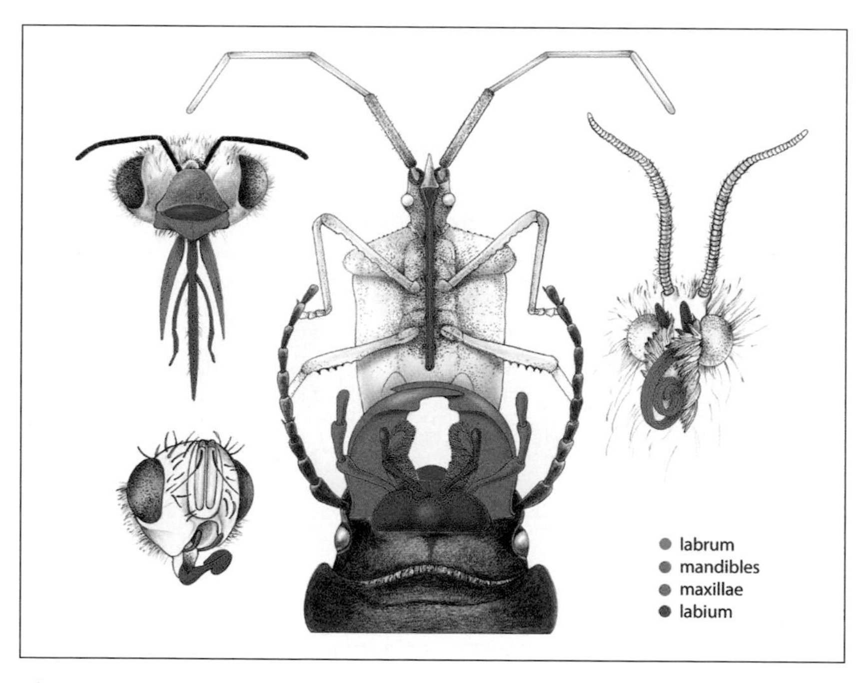

Plate 5.6. How insect mouthparts diverged. They can be sponge-like in a fly (lower left), suited for chewing and lapping in a bee (upper left), mandibulate for chewing in a beetle (center bottom), or tubular for sucking in a moth (right), or for piercing and sucking in a true bug (center top). Colors identify structures that are homologous, or shared due to common ancestry. (Drawing by Karen Anne Klein and Barrett Klein)

Plate 5.7. Stages on life's way: Division of labor among segments has allowed arthropods to diversify. The results include a shrimp, *Periclimenes yucataneus,* from Panama's Caribbean shore. (Photograph by Christian Ziegler)

Plate 5.8. A different specialization of segments yielded a praying mantis, *Creobater,* from Thailand. (Photograph by Christian Ziegler)

Plate 5.9. A rolled-up isopod from Borneo, representing another way of specializing segments. (Photograph by Christian Ziegler)

Plate 5.10. This mollusk is a chiton, *Mopalia* sp. from Tatoosh. (Photograph by Anne Paine)

Plate 5.11. This mollusk is a coral-eating snail, *Cyphoma gibbosa,* crawling on a "sea fan" or gorgonian, leaving a path of grazed tissue. (Photograph by J. L. Wulff)

Plate 5.12. These mollusks are bivalves, *Mytilus californianus,* from Tatoosh, which filter planktonic algae from the water. (Photograph by Anne Paine)

Plate 5.13. This mollusk is a poison ocellate octopus, *Amphioctopus mototi,* with a motile crinoid (an echinoderm) at Seraya Tulamben, Bali. (Photograph © Robert Delfs, 2014)

Plate 5.14. Zonation of animals and plants in the rocky intertidal at Tatoosh. Barnacles, *Balanus glandula,* form the cream-colored top zone, followed by an algal zone including *Mazzaella cornucopiae.* Below these algae is a mussel zone, below which is a zone dominated by the kelp *Hedophyllum sessile.* (Photograph by Anne Paine)

Plate 5.15. Different plants and animals dominate adjacent zones in the community inhabiting rocky intertidal shores at Tatoosh. Barnacles, *Balanus glandula,* are highest up, followed by mixed algae including *Mazzaella,* then the mussel zone, and finally the kelp zone, dominated by *Hedophyllum,* extending to the water's edge. The zonation repeats itself across the water. (Photograph by Anne Paine)

Plate 5.16. A starfish, *Pisaster ochraceus.* These starfish prevent mussels from spreading lower down and replacing the kelps and coralline algae in the zone below the mussels. (Photograph by Anne Paine)

Plate 5.17. The trade-off between fast growth and effective defense: Where protected from waves, herbivores such as these sea urchins, *Strongylocentrotus droebachiensis,* transform their habitat, creating an "urchin barren" paved with coralline algal crusts. (Photograph by Anne Paine)

Plate 5.18. The trade-off between fast growth and effective defense: These sea palms, *Postelsia palmaeformis,* grow fast where waves are strong enough to protect them from herbivores, so they need not pay for their own defense. These *Postelsia* are growing in a small gap the waves ripped from a bed of mussels. (Photograph by Anne Paine)

Plate 5.19. Ancestors of land plants? A filmy fast-growing intertidal green alga that grows at less wave-beaten sites at Tatoosh, the sea lettuce *Ulva lactuca,* with some *Halosaccion* and other algae. (Photograph by Anne Paine)

Plate 5.20. *Mazzaella cornucopiae,* a red alga (though golden-colored) that at Tatoosh forms a springing turf 2–3 centimeters high with 7 square centimeters of frond per square centimeter of rock. These algae often form a zone separating barnacles, *Balanus glandula,* from mussels at sites not fully exposed to the afternoon sun. (Photograph by Anne Paine)

Plate 5.21. Brown algae, including a stalked kelp, *Laminaria setchelli,* growing closest to the water, and *Hedophyllum sessile,* whose fronds are lying on the rock, just above it. (Photograph by Anne Paine)

Plate 5.22. A liverwort, *Marchantia.* Liverworts were the first plants to spread over the land. (Photograph by Christian Ziegler)

Plate 5.23. The alternation of generations between a leafy haploid gametophyte and a diploid, non-photosynthetic sporophyte bearing a spore capsule on a stalk, in a moss. (Photograph by Christian Ziegler)

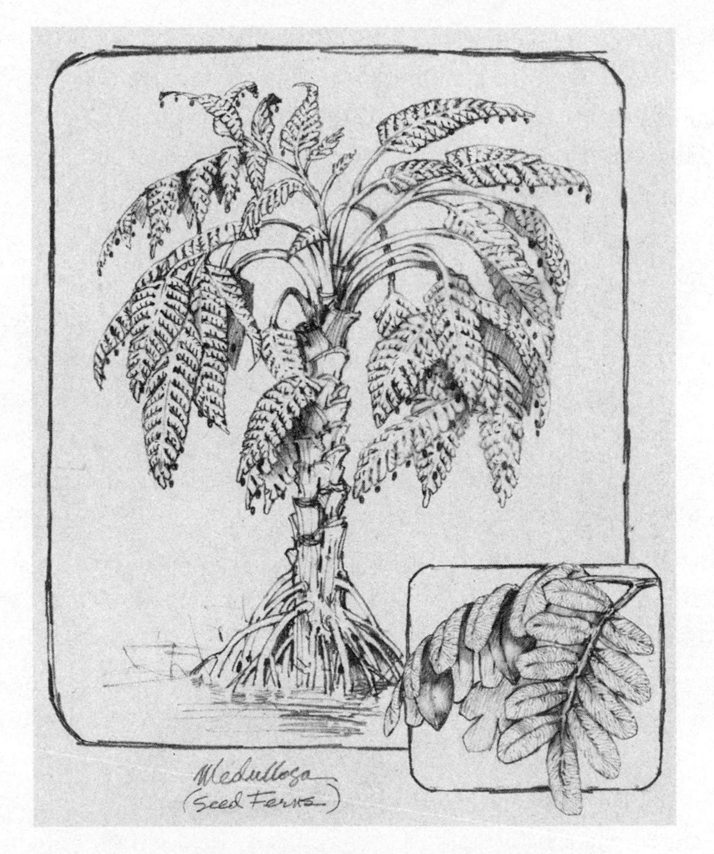

Fig. 5.16. A swamp forest seed fern, *Medullosa,* 5 meters tall. Seed ferns looked like tree ferns, but produced seeds, not spores, and were much eaten by arthropod herbivores. Both *Psaronius* and *Medullosa* lived in Carboniferous swamps 300 million years ago. (Drawing by Debby Cotter Kaspari)

"Coniferous forests of pine, spruce, fir, larch, cedar, cypress and juniper are the 'dark and gloomy forests' of Hiawatha, so unrewarding to animals and indeed to other plants."

THE EVOLUTION OF FLOWERING PLANTS

Flowering plants found a new way to cope with a world full of herbivores. They evolved flowers to attract animals that would convey pollen from one plant to another of its species. Dedicated pollinators, which focus on one or a few kinds of plants at a time, allow a plant species to maintain genetic variation even when a specialized pest keeps its individuals so rare and scattered that enough escape discovery to maintain the plant's numbers (plate 5.25). In the wet tropics, where plants are most diverse, plants escaping their specialist pests only need to defend themselves against less destructive generalist pests, so they can divert resources from defense to growing faster.

In the Jurassic, long before flowering plants evolved, cycads and other gymnosperms attracted insect pollinators, including butterfly-like lacewings unrelated to modern butterflies, with brightly colored wings and tubular mouths through which they sucked fluid. When a gymnosperm ovule is ready to be fertilized, the structure enclosing it secretes a "pollination drop" at its opening to snare appropriate pollen grains. This drop is often nutritious enough to attract insects. Those species of insects that preferred pollen or pollination drops from a single kind of plant when it was in season became pollinators; plants helped by placing pollen right by their pollination drops. Well before the Jurassic ended, members of a now extinct conifer family, some cycads, and a nearly extinct order of broad-leafed gymnosperms, Gnetales, attracted insect pollinators. Only the Gnetales ever seriously threatened the dominance of flowering plants. Though common beforehand, Gnetales nearly died out 90 million years ago; no one knows why. Some cycads have domesticated beetle pollinators that live and mate on their cones and only pollinate plants of their host species. Cycads, however, could not possibly outgrow flowering plants. The "pipes" bringing water from their roots to their leaves were narrow, short tracheids. Water, moreover, moved far more slowly from one tracheid to another than in modern conifers. Thus cycads transported water too slowly to grow fast. Today's slow-growing cycads are largely restricted to the understory of tropical and subtropical forests: they invest heavily in powerful anti-herbivore poisons.

In short, animal pollinators allow their plants to dominate the forest only if they, especially their seedlings, can grow fast. Just as effective systems of road, rail, and air transport allow faster economic growth, and effective networks for circulating blood allow animals to be more active and responsive, so plants that can move water readily from roots to and throughout their leaves can photosynthesize and grow fast. Unlike cycads, ancestral flowering plants were fast-growing weedy herbs of riverside forest.

The contrast between modern conifers and flowering plants reflects the trade-off between effective defense and fast growth. Growth is largely limited by photosynthetic capacity, the plant's maximum rate of photosynthesis per square millimeter of leaf. In turn, photosynthetic capacity is largely governed by how fast water can move through root, branch, and twig: the better its water transport system, the higher the photosynthetic capacity of its leaves can be. Conifers move water through small, narrow tracheids, but water can move far more rapidly between conifer than between cycad tracheids. Water also moves far more easily between conifer tracheids than between vessels of flowering trees, but in favorable habitats flowering trees compensate by making vessels four times wider and many times longer than conifer tracheids.

The advantage of flowering plants was greatly enhanced 100 million years ago by increasing the density of veins in their leaves. Today, the average flowering plant has 8 millimeters of veins per square millimeter of leaf, compared with 2 in the average conifer, or in the average angiosperm of 120 million years ago (plates 5.26, 5.27). A leaf's "vein density" (millimeters of vein per square millimeter of leaf) influences how fast water vapor can pass through its stomates, which in turn limits the leaf's photosynthetic capacity (how much sugar it can produce per square millimeter of leaf). Indeed, the vein density of a fossil leaf provides a (very crude) estimate of its photosynthetic capacity, showing how understanding modern plants and animals helps us understand fossil counterparts. Increased vein density increased transpiration, which tends to increase rainfall. In the tropics, the high transpiration of flowering forest created wetter, cooler climates with shorter dry seasons, tailoring the climate more nearly to its own needs. The takeover of tropical and many higher-latitude forests by flowering plants between 120 million and 70 million years ago made these forests much more hospitable to animals. As E. J. H. Corner remarked, "The flowering forests began to feed them, to use them, to foster them, and to bring forth their counterpart in the multitude of insects. . . . Animals do not help plant plankton, seaweeds, mosses, ferns and conifers; they help themselves to them. The flower symbolizes the new community, and indeed, *floreat!*"

Another water management problem is coping with dry season. In tropical trees, water management style falls between two extremes. Drought evaders store water in their trunks, which they release at need to keep tension low in their sap. If sap tension increases unduly they release a hormone that closes their stomates. In dry season they shed leaves and fine roots to cut water loss to dry, often hot and breezy, air and dry soil, and they regrow when rains return. Some trees have green, stomate-less strips on their trunks where photosynthesis recycles carbon dioxide respired by the bark. Drought tolerators have deep roots, and narrow thick-walled xylem vessels that slow sap flow. This reduces photosynthesis, but the vessels do not implode even under tension strong enough to pull water from deep roots or dry soil. Their stomates close when leaf water tension is too high. Temperate-zone trees face a similar trade-off: deciduous hardwoods photosynthesize abundantly in summer; evergreen conifers sacrifice some summer photosynthesis to function all year long.

Thanks to their lower photosynthetic capacity, conifer seedlings get a much slower start in life than their flowering counterparts. Thanks to being able to move water quickly, flowering trees dominate favorable habitats: lowland tropical forest, and most temperate-zone forests. Conifers dominate boreal forest, temperate-zone forests on poor soil, and many mountaintops in the Old World tropics. Evolving effective systems for transporting water, the latest stage in freeing plants from the shackles of their aquatic heritage, allowed flowering plants to use animal pollinators to create diverse, productive forest in favorable habitats, and to expand the area suitable for their growth. Indeed, animal pollination of pest-plagued, fast-growing flowering plants allowed terrestrial diversity of species visible to the naked eye to far outpass marine diversity. The land now has ten times as many macroscopic species as the sea, whereas before flowering plants evolved the numbers were nearly equal.

In sum, plants took full advantage of life on land by improving ways of bringing water and nutrients from soil to leaves and carbohydrates from leaves to roots. Besides effective internal transport systems and tall, strong, slender trunks to lift their leaves beyond competitors' shade, they developed pollen and seeds for more effective reproduction. Employing animals as pollinators and seed dispersers allowed flowering plants to reduce the costs of resisting pests and pathogens. Increasing the density of veins in their leaves further increased photosynthesis, so flowering plants dominated most forests. Their higher transpiration created wetter, cooler climates, and extended the range of rain forest—one more case when evolution in living beings made the earth a better home for life. The climax of these developments is tropical rain forest, where competition is most intense, life most diverse, and mutualism most prevalent, and most intricate.

SIX

Integrating Diversity into Community

Interdependence and Mutualism

ALL LIVING THINGS, INCLUDING HUMAN beings, compete for the means to live and reproduce. A few avoid competitors by making livings in new ways—tapping new energy sources, using others' wastes in new ways, or colonizing new habitats. Such stratagems open more occupations, more ways to make livings, in both human economies and natural ecosystems. How is this diversity integrated into ecological communities? How does competition among a community's members enhance interdependence and mutualism? How do interdependence and mutualism increase a community's productivity? What factors favor, and what barriers hinder, productivity increase?

Human Economies: How Competition Favors Diversification and Interdependence

To understand why living things have diversified, let us first ask why there are so many human occupations. A traditional band of hunter-gatherers, such as all human beings lived in 20,000 years ago, offers few alternatives for making a living. During the past 10,000 years, however, innovations, starting with agriculture, have decreased the fraction of people devoted to growing or gathering food, thus increasing diversity of human occupations, productivity of human economies, and interdependence among human communities. Moreover, this interdependence has extended over ever widening scales. The key step was agriculture, which produced enough to support people in other occupations. In favorable settings, like Mesopotamia and the north China plain, agriculture prompted new technologies, such as irrigation, and new tools, such as plows, scythes, and wheelbarrows, which increased productivity and opened new occupations for their makers and users, but also increased interdependence. Improved agriculture let fewer people grow more food that was easier to store and transport, freeing others for different occupations. These others, however, now depended on farmers.

Surplus wealth, created by improved agriculture, supported further innovation. A new mode of information transfer, writing, created livings for scribes, teachers—and historians such as Thucydides and Ssu-ma Ch'ien, philosophers such as Aristotle and Chuang Tzu, and mathematicians such as Euclid. Writing helped retain and spread knowledge and enabled better-coordinated cooperation. Printing amplified these developments manyfold, as has electronic communication.

Why so many occupations? No one person can do all things well. In a village of farmers, self-interest can spontaneously create a division of labor, which in turn increases interdependence. If enough farmers are within reach, someone particularly apt at working with wood may make tools and furniture in return for food. If a town is within reach of the transport available, another may live as a trader, bringing, for suitable reward, food to townspeople and products made by these townsmen to farmers who want them. This division of labor, this specialization of occupations, is driven by the search for ways to make better livings. It enhances output by reducing the time taken to switch tasks and adjust to new tasks requiring different skills. It also heightens interdependence. In 1776, Adam Smith could already write: "The woollen coat . . . which covers the day-labourer . . . is the produce of the joint labour of a great multitude of workmen. The shepherd, the sorter of the wool, the wool-comber or carder, the dyer, the scribbler, the spinner, the weaver, the fuller, the dresser, with many others, must all join their different arts in order to complete even this homely production. How many merchants and carriers, besides, must have been employed in transporting the materials from some of these workmen to others who often live in a very distant part of the country! How much commerce and navigation in particular, how many ship-builders, sailors, sail-makers, rope-makers, must have been employed in order to bring together the different drugs made use of by the dyer." He also showed how more widespread wealth and easier transport increases the diversity of feasible occupations, by expanding the field of customers enough to support specialized craftsmen like goldsmiths or jewellers.

Division of labor often reflects planned cooperation. Adam Smith began his *Enquiry into the Nature and Causes of the Wealth of Nations* by showing how division of labor, partitioning an enterprise's work into a set of specialized tasks, each performed by different people, multiplies output per person. Making pins, for example, involves drawing out the wire from which they are made, straightening it, cutting it, sharpening the pinpoints, preparing their other ends for the attachment of pinheads, making these pinheads, putting them on the pins, and so forth. Unaided, a person cannot make more than twenty pins a day; there are too many different things to do, and one worker will never have time to become skilled at them all, or enough money to buy machinery to do each task more quickly and easily. On the other hand, a company of ten workers, each devoted to a different set of tasks, can make 48,000 pins per day—4,800 per person. This division of labor reflects deliberate cooperation to make pins.

Competition drives economic development. Coal-powered and, later, oil-powered steamships replaced sailing ships for many but not all purposes; in poor countries, sail still powers small fishing boats. Similarly, railway trains and trucks replaced animal-drawn wagons for bulk freight hauled long distances, although in less-developed countries, and some districts of wealthier countries, animals still power freight transport for some shorter hauls. Similarly, airplanes replaced trains and trucks for rapid transport of valuable or perishable cargo; passenger trains, automobiles, and later airplanes replaced horses and horse-drawn carriages; and telegraph, telephone, fax, and internet largely replaced postal and messenger services. Greater speed and wider scale of communication and transport make possible ever more intricate cooperation on ever wider scales. New sources of power, and new modes of manufacture, make such cooperation profitable. These changes increased variety of occupations, degree of specialization, and economic productivity per capita.

In short, improved communication and transport enable more elaborate forms of cooperation and coordination of workers with more different skills in larger, more complex business firms. Building metal steamships needs more different resources and coordinates more different kinds of work than building wooden sailing ships. Building railway locomotives demands many more skills, and many more different machines, than making a freight wagon. Moreover, such locomotives require tracks to run on, and wood, coal, or oil to run at all; they demand many new forms of coordination. Forming ever larger, better coordinated companies to promote innovations diversifies occupations and generates more wealth. Such economy-transforming innovations often require large pools of labor, many skilled workers, and abundant capital. Innovations such as steamships, automobiles and highways, railway systems, and airplane service first appeared in large, productive economies.

However, some factors hinder production. An individual or group that robs, enslaves, or kills competitors competes unfairly. To benefit society, competition must be "fair": it must not impair the common good. An individual or group produces more if it benefits differentially from its own work, but that is not enough. Monopolies could still disrupt the common good by denying resources to those who can use them best. In large, wealthy societies, moreover, the powerful can bend rules of competition in their favor, which often retards productivity. To keep competition fair, a society's members must be willing and able to discern and enforce fairness, either vicariously through agents of the law, or directly, as in hunter-gatherer bands.

The productivity of human economies hinges on cooperative enterprise where individuals pool different skills to exploit new opportunities or to compete better with others. The many laws against cheating in cooperative ventures—violating contracts, stealing profits of cooperation from their partners, and the like, highlight the crucial importance to society of cooperative enterprise.

Can human economies show us what factors intensify competition and favor diversification in natural ecosystems? How is unfair competition restrained in these systems? How does interdependence enhance productivity?

Productivity and Its Hindrances

A random change in the design of a machine organized for a function usually makes it less apt, if not totally unfit, for its function. Similarly, ecosystems, natural economies, are organized to maintain high productivity and diversity, because both drop when their environments are altered in unprecedented ways. Novel environmental changes induced by human activity, whether killing crucial predators or seed dispersers, fragmenting habitats, or warming the planet, usually reduce productivity, diversity, or both. Abandoning farmland cleared from rain forest can let sterile grassland, depressingly unlike the luxuriant, diverse forest that once grew there, take over (plate 6.1). Killing off the wolves and puma in the forests of eastern North America allowed the deer they formerly ate to increase to a level where only the worst tasting, least nutritious plants could survive. For several million years, sea otters protected the kelp forests surrounding the Aleutian Islands by eating sea urchins and other invertebrate herbivores, allowing the kelps to invest in fast growth rather than anti-herbivore defense. These fast-growing kelp forests harbored many animals. When hunters killed all these otters for their fur, the sea urchins they formerly ate multiplied, grazed down the productive kelp forests, and replaced them with "urchin barrens."

Further evidence that natural ecosystems are organized for high productivity and diversity is that both increased over evolutionary time. Finally, productivity and diversity always recover after mass extinctions, such as the one 65 million years ago that ended the Cretaceous, or the one 250 million years ago that ended the Permian.

Still, one factor often limits production. A "virgin" forest, with its soaring tree trunks lit by shafts of sunlight filtering through the leaves of its tall trees, is a scene of beauty—which reflects a "tragedy of the commons." Each tree that grows taller to benefit itself makes life worse for all. Leaves of shrubs rising above a field of herbs catch more light than their neighbors' do, so they reproduce faster, and the meadow becomes a shrubland, the first step toward a forest, where other plants must "keep up with the Joneses" to maintain their access to light. Forest trees must build tall and

Fig. 6.1. A tragedy of the commons. Each redwood tree must make huge amounts of wood to keep up with its neighbors and place its leaves in the sun, whereas grass in an open grassland need make no wood at all to do this. (Photograph by Tim Fitzharris/Minden Pictures)

stout woody trunks and big branches to give their leaves as good access to sunlight as their distant short-stemmed ancestors had "for free." A hectare of tropical forest 40 meters tall in Panama uses 281 tons dry weight of wood to support its 7 tons dry weight (7 hectares) of leaves, and spends more of its resources on making wood than making leaves. A hectare of redwoods 100 meters tall uses 4,100 tons of wood to support its 17 tons (8.4 hectares) of leaves (fig. 6.1).

Transpiration pulls water from soil to leaves. Lifting water to higher leaves increases its tension. If tension sucks air into a tracheid, breaking the water column, it cannot be refilled. Tall trees must therefore adjust partitions between tracheids to prevent the air in an embolized tracheid from being sucked into its neighbors, which reduces water flow. Because a leaf lets water out as vapor to let CO_2 in, reducing water flow reduces photosynthesis. This is why redwood leaves 50 meters above the ground have 1.8 times the photosynthetic capacity per gram of leaf as leaves at 80 meters, and 6 times that of leaves at 110 meters.

Finally, lifting a tree's leaves higher imposes productivity-reducing inequalities. In mature tropical forest with 7 hectares of leaves per hectare of ground, each hectare of leaves takes up half the remaining light. Only about 1 percent of the light above the canopy reaches the forest floor, barely enough to allow an herb growing on the forest floor to cope with herbivores and replace leaves as needed (plate 6.2). Light is therefore distributed very unequally. Although herbs on the forest floor get just enough light to survive, leaves atop tree crowns receive far more light than they can use (plate 6.3). Foresters know that a young forest's productivity drops when its canopy closes—when tree crowns join to form continuous "canopy," light is distributed far less evenly among that forest's leaves.

Favorable Climates Intensify Competition, Diversity, and Interdependence

As in human economies, so in natural ecosystems, abundant resources favor competition, innovation, diversification, and interdependence. Tropical inhabitants do not face a brutal alternation of cold, foodless winters and hot summers (plate 6.4). Like Darwin, the evolutionary biologist Theodosius Dobzhansky saw that whereas coping with hostile climate dominates life at higher latitudes, interactions with other living beings dominate life in the tropics, where organisms must compete with each other, or help each other compete better with third parties, to survive. Tropical climates, moreover, allow higher production of vegetable matter. In rain forests of Amazonia or Costa Rica, photosynthesis produces over twice as much sugar (carbohydrate) as in Harvard Forest (Table 1), where deciduous trees are fully leafed out only from May through September.

In tropical settings, intense competition ensures ruthless exploitation of all possible energy sources. Wooden houses, clothes, and books are a magnet for termites and other insects. Geckoes multiply by eating house-

Table 1. Total photosynthesis, expressed as tons of sugar (carbohydrates) photosynthesized per hectare per year, in different forests

Latitude	*Site*	*Total photosynthesis*
43° N.	Harvard Forest, Massachusetts	30
10° N.	Rain forest, La Selva, Costa Rica	74
2° S.	Rain forest, central Amazonia	76

hold insects. Even camera lenses attract fungi that etch them. Especially in the tropics, monopolies of poorly used resources rarely last long: energy sources usually wind up attracting efficient users.

On good soil, tropical plants compete especially intensely for light. The contrast between the open understory of intact tropical forest and the dense regrowth where the fall of a great tree lets abundant light reach the ground is particularly striking. Moreover, tropical forest houses many plants like bird-nest ferns and bromeliads that perch on tree branches, and lianas, thick woody vines that climb trees to reach a place in the sun, rather than building tree trunks to support their own leaves (plate 6.5).

Tropical plants also need pollinators. Tropical forest lacks the massive and brief summertime floral displays of alpine meadows, because winter's absence allows tropical plants to flower at any time of year. Nonetheless, the proverbial splendor and sweetness of some tropical flowers, and the many bizarre ways less sweet smelling flowers attract pollinators, reveal how intensely tropical plants compete for pollinators (plate 6.6).

Indeed, intense competition favors interdependence. Many animals must exploit several habitats. In Panama's forests, damselfly nymphs and mosquito wrigglers grow in water-filled tree holes, while their adults range throughout the forest (plate 6.7). Most tadpoles live in ponds or streams, their parent frogs on land. Different regions are interdependent. Many birds migrate to high latitudes to breed, return to the tropics to escape winter, and disperse seeds in both places. In East Africa's Serengeti grasslands, a huge herd of wildebeest follows the rains from place to place, cropping the fresh grass these rains bring forth. Many tropical butterflies also follow the rains, laying eggs on the tender leaves that flush then (plate 6.8). In tropical seas, mangroves are "nurseries" for fish whose adults live on coral reefs. Tropical communities are webs of interdependence.

Tropical animals compete intensely for food. To cope with competitors, each species specializes to a distinct food or combination of foods, or to food gotten by a distinct technique. Higher diversity within a guild of insect-eaters reflects more intense competition among them. Old-growth forest in South Carolina harbors 25 species of bird that live primarily on insects, whereas an equivalent area of mildly seasonal tropical forest in southeast Peru harbors 95, and everwet rain forest in French Guiana, about 100.

A predator competes for the resources in its prey's bodies. Predator pressure is most intense in the tropics, as evidenced by the wondrous variety of ways tropical insects manage to look like sticks, leaves, thorns, bird droppings, or more poisonous insects. High predator pressure is likewise reflected by the garish, attention-grabbing brightness of the colors by which other insects, and even frogs, advertise their distastefulness.

Herbivory and disease are also most devastating in the tropics. Winter cold keeps most insects from moving or eating and checks disease-causing microbes and fungi. Their populations must grow anew every year, whereas tropical pests—insects and disease—can work all year long, intensifying pest pressure in tropical habitats. Mature leaves of most tropical trees are too tough to eat. On the average, however, young leaves in Panama are eaten several times more rapidly, despite being much more poisonous, than young temperate-zone hardwood leaves. In the tropics, tender young leaves are defended by chemicals that usually prevent serious damage by all except a few specialists adapted to handle that defense. Some tropical trees, like Panama's *Gustavia superba,* expand and toughen their leaves as fast as possible. *Gustavia*-eating caterpillars survive only if they hatch on leaves less than eight days old. Other plants supplement young leaves' chemical defenses by using sweet liquid in "nectaries" on their leaves to attract frightening ants. Other trees feed and house ants that survive only by successfully defending their hosts. Plants with slow-growing young leaves stuff them with poisons. "Bioprospectors" test such leaves for disease-curing chemi-

Table 2. Number of trees, and number of tree species, on 1-hectare forest plots in the tropics and the north temperate zone

Latitude	*Site*	*# of trees*	*# of tree species*
1° S.	Everwet old-growth forest, Ecuador	654	239
12° S.	Mildly seasonal old growth, SE Peru	598	174
9° N.	Seasonal old-growth forest, Panama	409	91
11° N.	Dry forest, Costa Rica	354	56
42° N.	Mixed old-growth forest, Manchuria	421	16

cals. No one defense, however, parries all pests, nor can any pest penetrate all defenses. A specialist pest can devastate overly dense host populations, just as epidemics devastate crowded refugee camps. Do specialist tropical pests keep their host so rare that, amid all the foliage they cannot attack, few of these pests find enough leaves of the right kind to survive?

In fact, most tropical forests are diverse because each species is somehow kept rare, making room for many others. Seedlings establish least often, grow slowest, and die fastest where more members of their species are nearby. Do specialist pests drive these patterns? In Manchuria, where winter is severe, a hectare of old-growth forest has 16 tree species; on Barro Colorado Island, Panama, where no winter checks pest populations, a hectare has about 90 tree species (Table 2). Like winter, but to a lesser degree, severe dry seasons reduce pest pressure, lowering tree diversity. In south Indian forest, where the dry season can last six months, trees reduce pest damage by leafing out before the rains come. Many trees do likewise on Barro Colorado, where dry season lasts four months. A hectare of everwet old growth in Ecuador, where no dry season checks its pests, has about 240 tree species, far more than Barro Colorado, where only 5 percent of the year's 2.6 meters of rain falls between January 1 and March 31. In diverse tree genera, closely related species differ primarily in their anti-pest defenses, as if avoiding a close competitor's pests were the best way to coexist with it. Specialist pests thus appear to enforce tropical tree diversity. Which pests do this? Attention long focused on insects. They and their damage are easily seen, especially when an insect's numbers explode. Insect outbreaks usually afflict one or a few related plant species; the most damaging pests are specialists.

Despite insects' obvious impact, spraying insecticide on square-meter plots in tropical forest of Belize reduces seedling diversity only slightly, although it changes the species composition of emerging seedlings. Do insect-eaters limit insect pests? On Barro Colorado, insects annually eat at least 500 kilograms dry weight of leaves per hectare—7 percent of the supply. Birds eat about 20 kilograms dry weight of these insects. If it takes 10 kilograms of leaves to make a kilogram of insects, perhaps a third of the foliage insects eat feeds insects birds eat. Bats eat more leaf-eating insects than birds (plate 6.9). A graduate student, Margareta Kalka, found that excluding bats from plants at night damaged plants more than excluding birds from plants of the same species by day.

Experiments by a postdoctoral fellow, Scott Mangan, in Panama show that seedlings grow slower and die faster in soil from under adults of their species than from soil from under others. Rarer species in natural forest appear to be those whose seedlings suffer more from being grown in soil from under adults of their species. Soil microbes may keep each tree species rare enough to make room for many others.

In sum, intense competition favors a great diversity of tropical life, and makes tropical forests more productive. Tropical specialists exploit resources more fully than counterparts at higher latitudes. More intense herbivory and predation cycle resources more rapidly, fueling faster growth, a quicker pace of life, and higher productivity.

As evolution progressed, plants everywhere faced ever more kinds of herbivores and diseases, animals faced a growing variety of increasingly active predators, yet productivity increased, especially in the tropics, where these pressures are greatest. As more wealth lets one replace an efficient Volkswagen by a speedy, gas-guzzling Ferrari, so higher productivity allowed competitive dominants, the principal herbivores and carnivores, to be replaced by less efficient but more active, responsive successors with higher metabolism and relatively larger brains. Effective competitors, such as flowering plants and resource-demanding human beings, usually appeared first in the tropics and spread later to higher latitudes. Indeed, at higher latitudes, many flowering plants, like oaks, beeches, and maples, revert to wind pollination: are animal pollinators scarce thanks to winter, or are these trees too common to need animal pollinators?

Cooperating the Better to Compete

Especially in the tropics, the need to compete effectively favored many forms of cooperation. By cooperating, individuals can compete better with others, just as several boys can jointly fend off a playground bully that could defeat any one alone. Tropical forests shelter many insect societies whose elaborate social organizations have few parallels in the temperate zone. Insect societies changed the face of natural economies, especially in the tropics. Termite societies broke down wood, and in some areas, leaf litter, far more rapidly, speeding the recycling of mineral nutrients tied up in dead matter. Social life opened many roles for ants, especially in tropical forest. The massive swarm raids of army ants and driver ants, and leaf-cutter ant colonies with millions of workers, characterize many tropical habitats. Bees became principal pollinators of flowering plants (plate 6.10). Some evolved complex societies. Social life, however, requires elaborate sensory, locomotory, navigational, and cognitive capacities and effective signaling systems to partition tasks among, and coordinate activities of, the colony's workers. Among bees with truly complex societies, only European honeybees mastered the difficult art of surviving winter. When queens of African honeybees introduced into South America escaped from their cages, however, their "Africanized" progeny spread like wildfire throughout the lowland Neotropics, replacing lowland descendants of the honeybees European colonists imported to make honey.

Some tropical innovations hinge on spectacular cooperation between members of different species. Leaf-cutter ants became effective herbivores by domesticating a fungus that can digest many different kinds of leaves, making leaf-cutters major consumers of tree leaves in many Neotropical forests, where they eat 3 percent or more of the total leaf production. Fungus-growing termites became the principal decomposers of dead vegetable matter in tropical East Africa by cultivating a fungus that digested wood, dry grass, or leaf litter that termites bring in from outside the nest. Like leaf-cutter ants, these termites build nests of several cubic meters housing millions of workers.

MUTUALISMS BETWEEN TREES AND THEIR POLLINATORS AND SEED DISPERSERS

As animals became more active and discriminating, and better at moving long distances and finding their way, cooperation between plants and animals helped defend tropical forests against pests and diseases. Conspicuous flowers attracting wide-ranging animal pollinators that temporarily focus on plants of a single species enabled even plants of rare species to receive pollen from other plants of their kind. Thus, as the previous chapter showed, a plant species rare enough that most of its individuals escape their specialist pests can maintain enough genetic variation to resist disease and shift resources from anti-herbivore defense to growing faster. Thus animal pollinators — and fruit-eating seed dispersers — enabled a diverse array of rare, fast-growing flowering plants to replace a less diverse set of slower-growing, better-defended, wind-pollinated conifers. In tropical forest, where herbivory is most intense, flowering plants spread because animal pollinators and dispersers allowed them to escape specialist pests and outgrow their competition. The spread of faster-growing, shorter-lived flowering trees increased forest productivity and resource turnover, and diversity of trees. Cooperation between animals and flowering plants transformed forest ecosystems, making them more mutualistic. A flowering tree, however, is far less "self-sufficient" than a conifer: it needs not only its pollinators and seed-dispersers, but also trees of other species, perhaps even in other forests, to feed these animals when it has no flowers or fruit.

Pollinators provide so important a service that some plants, such as fig trees, "domesticate" them. Each species of fig tree maintains one or more species of pollinating wasp. A fig fruit, called a syconium, begins as a flowerhead turned outside in. Its flowers line the inside of a ball with a hole at one end. When fig flowers are ready to pollinate, their odor attracts one or more fertilized, pollen-bearing female wasps, the "foundresses," to enter the syconium and pollinate the flowers. In half of these flowers, a wasp stings the ovule (incipient seed) at its base, transforming it into a proper home for a wasp larva, and lays an egg in it. When these larvae mature, they mate with each other, and the mated females, carrying pollen, leave their syconium (plate 6.11). A fig tree ripens all its syconia at once, so emerging wasps must find other trees of its species with flowers ready to pollinate. Although fig wasps are minute, and live less than three days, they ascend above the canopy and drift in the breeze until they smell a receptive fig of the right species. They can pollinate trees over 10 kilometers from where they were born. These far-ranging pollinators maintain unusually high genetic variability even in rare species with less than one adult tree per 10 hectares. Fig trees, the "ultimate" flowering trees, are the fastest-growing and most nutritious in the forest. As E. J. H. Corner said, "By leaf, fruit, and easily rotted wood fig-plants supply an abundance of surplus produce."

"Domesticating" these pollinator wasps, rendering them totally dependent on the fig trees they pollinate, is effective, but it costs. Each species of fig must have trees in fruit at all times of year, regardless of when fig seedlings grow best, so that these short-lived wasps can find trees ready to receive their eggs and pollen before they die. Barro Colorado Island's 17 species of fig therefore produce a steady supply of figs that supports 10 bat species, each deriving half or more of its food from fig fruit.

Once flowering plants took over tropical forests, about 65 million years ago, fruit-eating and nectar-drinking mammals, including monkeys and bats; birds such as guans, toucans, hornbills, tanagers, and hummingbirds; and above all, insects, diversified greatly in response to this new opportunity. Living on nectar, pollen, or fruit is not easy: clumps of flowers and fruit of different sizes appear in different places, at different times, lasting for different periods. Social insect pollinators evolved spectacular techniques for telling their fellow workers the location and quality of flowers, and coordinating these workers' activities. In social primate groups, moving through and between tree crowns, tracking fruit sources, coordinating access to food and responses to predators, and maintaining social harmony favored developments such as hand-eye coordination, and cognitive skills such as recognizing, and inferring intentions of, and relationships between, their group's members. For primates, these were crucial steps toward evolving intelligence. The productivity of natural ecosystems depends as much on mutualism among species as the productivity of human economies depends on cooperative enterprise.

MUTUALISM AND THE SPREAD OF CORAL REEFS

The history of reefs is wonderfully discontinuous. Reef-building episodes can be separated by many million years without reefs. Do conditions favoring effective defense alternate with those favoring fast growth? Now, reefs flourish on hard bottoms in nutrient-poor water—stable, well-lit habitats (plate 6.12). The dominant animals are corals, clonal colonies of polyps—miniature sea anemones—each set in a limestone cup that deters predators. Polyps are organized to exchange food with each other. Corals owe their success to the symbiotic algae, zooxanthellae, they harbor. These algae provide 90 percent of the polyps' food in return for the nutrient-rich digestive wastes from minute animals that polyps snare, and accelerate the growth of the coral's limestone skeleton at least tenfold.

In the Indo-Pacific, other animals also maintain photosynthetic algae. *Tridacna,* giant clams (the largest clams living), nestle among reef corals, exposing brightly colored tissue, full of zooxanthellae, to the light. Energy from their algae allows them to drive powerful currents that suck into their bodies large zooplankton such as copepods: wastes from digesting these zooplankton presumably fertilize the algae. Their zooxanthellae apparently enable shells of giant clams to grow with unparalleled speed. After corals, sponges are the largest source of living matter on most coral reefs. Sponges often maintain symbiotic cyanobacteria. The Great Barrier Reef extends far eastward from Australia. Farther offshore, the water is clearer and poorer in nutrients and microbes. Many sponges there are shaped like flat dishes positioned to catch the light, and are full of cyanobacteria that supply most of their energy. Some of these sponges pass their bacteria from parent to offspring, as one does heirlooms.

Today, corals with zooxanthellae form more closely integrated colonies where living tissue connects the polyps, which are smaller. Sponges, clams, and corals with symbiotic algae are positioned so as to expose algae-laden tissues to sunlight. These characteristics are used to judge the presence of symbiotic algae in fossil counterparts.

Coral reefs were building up in the Silurian, 450 million years ago. In the Devonian, 80 million years later, tropical climates reached far higher latitudes than today, and reefs were larger and more widespread than ever after. The main reef-builders were tabulate corals and stromatoporoid sponges. A tabulate coral is built like a miniature block of one-room apartments, with new floors being added on top. Each top-floor apartment is roofless and harbors a polyp; lower apartments are empty. As reef-building progressed, polyps became more closely integrated. A stromatoporoid sponge is a thin layer of living tissue, organized like an encrusting sponge, atop the limestone layers it laid down before (fig. 6.2). Water enters its flagellar chambers through many small pores and flows from these chambers along canals converging on the nearest osculum: traces of these canals and oscula persist on the limestone surface. Oscula protruded above the sponge surface, making the water moving past the sponge help move water through the sponge. Tabulate corals and stromatoporoid sponges seem designed to be lit by the sun, as if they harbored symbiotic algae. In the Devonian, new kinds of fish evolved that ate hard-shelled prey, favoring massive, massively defended organisms over faster-growing competitors. Devonian reefs were full of "crypts" and "caves" housing filter-feeding calcareous sponges, nonsymbiotic rugose corals, and much else (fig. 6.3). Near the end of the Devonian, however, a worldwide mass extinction, probably caused by sudden, severe global warming, swept these reefs away, opening opportunities for new types of animals and plants to evolve.

In the Permian, a massive reef, "El Capitan," developed about 270 million years ago in what is now west Texas. It was composed of calcareous algae, large colonies of bryozoans, calcareous "sphinctozoan" sponges, and immobile brachiopods, filter-feeding animals with clamlike shells, affixed to their substrate, with lower metabolism than clams of their size. The largest brachiopods, principally of the family Richtofeniidae, were inge-

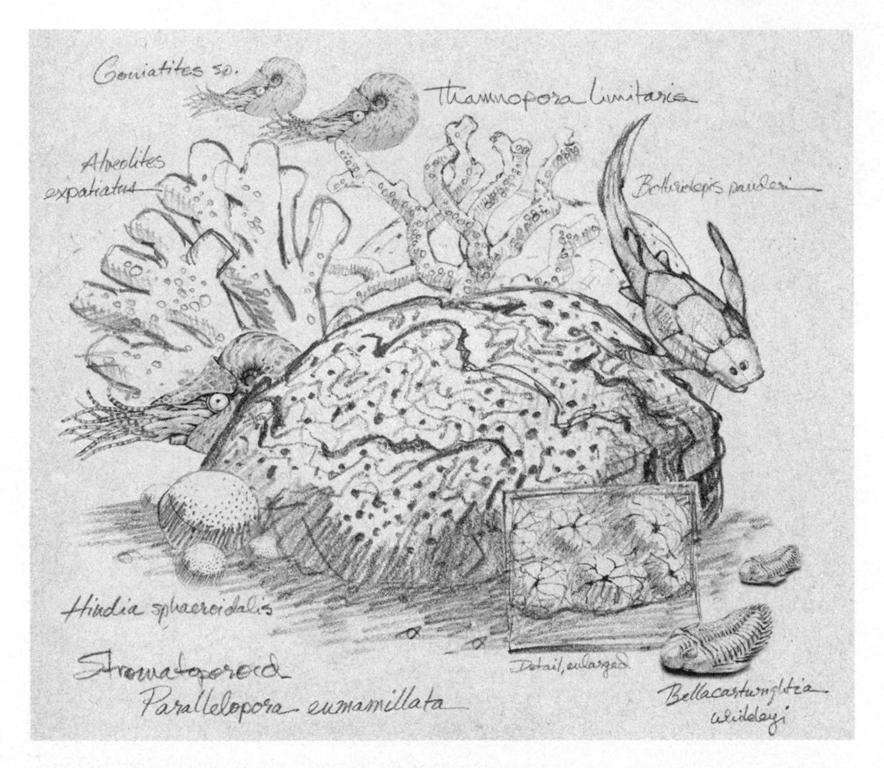

Fig. 6.2. A calcareous Paleozoic stromatoporoid sponge, *Parallelopora.* Such sponges lived on coral reefs: they may have harbored symbiotic algae. The *Goniatites* at upper left are nautiloid cephalopods, the *Thamnopora* and *Alveolites* behind the large stromatoporoid *Parallelopora* are tabulate corals, the armored fish at upper right is *Bothriolepis panderi,* the *Hindia* at lower left are smaller stromatoporoids, and the animals at lower right are trilobites, *Bellacartwrightia whiteleyi.* Inset shows the transport network on the surface of the *Parallelopora* "skeleton." (Drawing by Debby Cotter Kaspari)

niously designed to culture photosynthetic algae (fig. 6.4). They were vase-shaped, with a lid sunk well down in the vase: the lid and surrounding rim supported soft tissues. On coral reefs, with their many predators, such tissues are exposed almost always for the purpose of exposing symbiotic algae to the sun. The lid was easily lifted: it is thought that energy from the algae was used to open and shut the lid so rapidly that it sucked in large zooplankton, the wastes from digesting which fertilized the algae.

At the Permian's end, 251 million years ago, volcanic activity in Siberia on a scale seldom equaled extruded 5 million cubic kilometers — 1.2 million cubic miles — of lava, enough to cover all of Amazonia with a kilometer-thick layer of lava. Much of this erupted through coal seams, spewing huge quantities of carbon dioxide and methane into the atmosphere. Tropical oceans warmed from 27° C (82° F) to 36° C (96° F), killing these reefs, and 92 percent of the world's marine genera. Cyanobacteria, which few things ate, became the sea's primary producers. They were too small to sink fast, so the oxygen organisms that consumed to decompose them rendered the oceans anoxic and sulphidic. Near shore, cyanobacteria formed stromatolites. As in western Australia's Shark Bay today, the consumers that had kept stromatolites rare after the Precambrian could not live in such hot water. Land animals and plants also died out: this was the worst extinction of the past half-billion years. Most land plants died, providing an unparalleled feast for fungi. The tropics stayed hot enough for several million years to banish fish and trees to higher latitudes. Recovery was slow.

After the earth cooled again, the stony corals that build modern reefs evolved, and began cultivating zooxanthellae and forming reefs about 215 million years ago. When the Triassic ended, 200 million years ago, new lavas erupted, the earth warmed again, the seas again turned anoxic and sulphidic, plant productivity collapsed, and a new extinction wiped out these reefs and much else besides. Eventually, coral reefs recovered. A hundred million years ago, in the mid-Cretaceous, water again became warmer than today. Clams attached to the sea bottom by their lower valve evolved into "rudists," whose attached lower valve formed a long upright narrow vase; their upper valve was its lid, designed to support symbiotic algae (fig. 6.5). The rudists first coexisted with, then replaced, reef-building corals. A mass extinction 65 million years ago wiped out the rudists, along with dinosaurs and much else. From the origin of stony corals until the rudists displaced them, coral colonies became progressively more integrated. A few million years after this extinction, coral reefs like those we know today reappeared.

The earth's biota, however, coped better with later massive lava flows. The separation of Greenland from Britain began 56 million years ago with the extrusion through coal seams of 6 million cubic kilometers of lava.

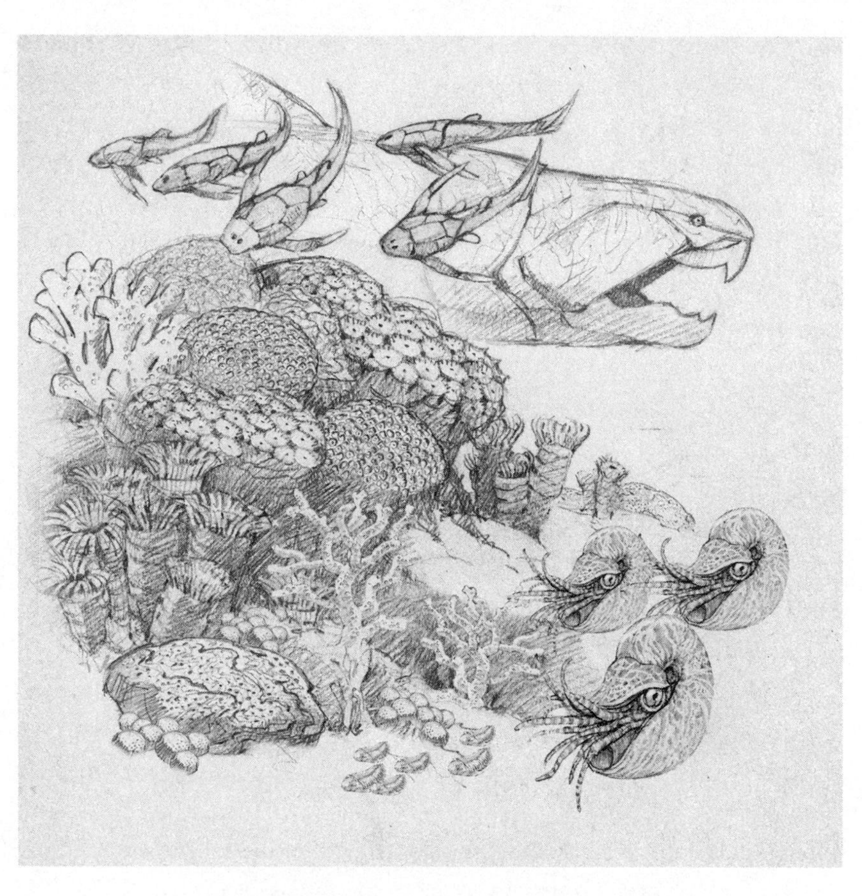

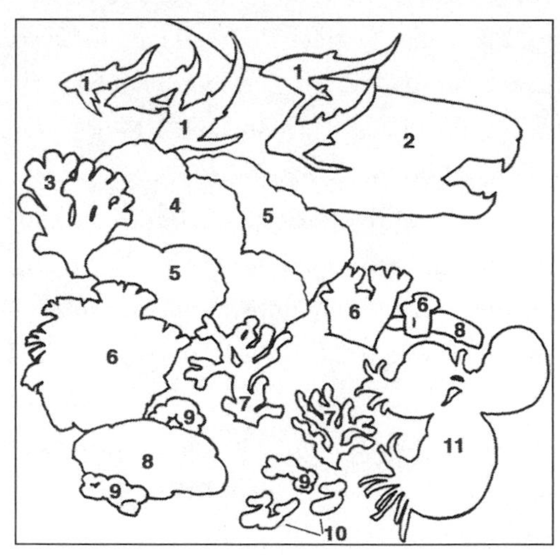
1
1
1
2
3
4
5
5
6
6
6
8
7
7
9
11
8
9
9
10

Fig. 6.4. Drawing of a group of richthofeniids, a reef brachiopod that maintained symbiotic photosynthetic algae. Their upper valve is a revolving lid that brings water and large zooplankton into the brachiopod, where it filters them out and eats them. The spotted tissues on the upper surface represent tissues containing zooxanthellae. (Drawing by Debby Cotter Kaspari)

Fig. 6.3. (*opposite*) A scene from a Devonian reef, made up largely of corals. The Devonian was the golden age of coral reefs, when reefs were far more extensive and magnificent than their modern counterparts. Shown here are (1) a detritus-eating fish with an armored head, *Bothriolepis panderi;* (2) a 4-ton armored fish, *Dunkleosteus terrilli,* this reef's top predator, 10 meters long; (3) a tabulate coral, *Alveolites expatiates;* (4) another tabulate coral, *Favosites mundus;* (5) a rugose coral, *Taimyrophyllum nolani;* (6) another rugose coral, *Heliophyllum vertical;* (7) a third tabulate coral, *Thamnopora limitaris;* (8) a stromatoporoid sponge, *Parallelopora eumamillata;* (9) another stromatoporoid sponge, *Hindia sphaeroidalis;* (10) trilobites, *Bellacartwrightia whiteleyi;* (11) nautiloid cephalopods, *Goniatites* sp. (Drawing by Debby Cotter Kaspari)

Fig. 6.5. The rudist, *Durania* sp., a clam modified to maintain photosynthetic algae on its cap, that took over reefs when the water was too hot for reef corals, at a time when dinosaurs still lived. (Drawing by Debby Cotter Kaspari)

Global temperature increased by 5°–8° C in less than 10,000 years, but, far from causing a mass extinction, it increased forest diversity and pressure from insect pests in the Dakotas and facilitated exchange of mammals between America and both Europe via Greenland and Asia via Alaska.

TURNING HERBIVORES INTO MUTUALISTS: THE COEVOLUTION OF GRASSLAND AND GRAZERS

Collusion between plants and their consumers caused grassland to evolve. A forest's trees are organized to resist herbivore damage. The wood of living trees is not very nutritious. Few animals have teeth, jaws, or digestive systems that can cope with it. Indeed, termite guts maintain a whole zoo of one-celled animals, protozoans equipped with digestive-enhancing live-in microbes that confer a bizarre variety of digestive abilities, to help digest the dead wood termites eat. Before organisms evolved that could eat it, lots of dead wood accumulated and became coal.

Many more animals eat leaves. A tough, full-grown leaf, however, is indigestible. Most leaf eaters prefer young leaves, which are tender and nutritious. In Panama, where most trees are evergreen, herbivores inflict 70 percent of the average leaf's lifetime damage in the month needed for it to attain full size and become too tough to eat. These leaves live long enough for their photosynthesis to repay the costs of becoming so tough, but in deciduous forests with growing seasons of seven months or less, leaves are too short-lived to repay these costs. In the temperate zone, 70 percent of the average lifetime damage is inflicted on a deciduous leaf after it attains full size. In south India, where large mammals like elephants survived human hunting, dry forests support several times more weight of mammals per square kilometer than their wetter, less seasonal counterparts. The weak defenses of tropical dry forest leaves led to the evolution of grassland, where herbivores excluded seedlings and saplings of competing trees. Now, grasslands differ radically from forests in how they handle herbivores. Broadly speaking, forests evolved to minimize herbivory, but grasslands evolved to support enough herbivores to exclude competing trees. Why?

Grasses evolved long before grasslands did, over 70 million years ago. Like early flowering plants, grasses got their start by evolving a new way of dealing with herbivores, only grasses perfected defense, not escape. Grass leaves grew from the ground-level tips of shoots sprouting from underground runners: their narrow, silica-rich leaves were difficult and unpleasant for herbivores to eat.

Gregory Retallack, a professor at the University of Oregon who studies fossil soils, showed how grasslands and grazers in the drier parts of Oregon, Montana, and Nebraska coevolved during the past 40 million years. Fossil soils are preserved in "badlands," bare hillocks of many-colored strata, each stratum representing a different soil (plate 6.13). These soils reveal much about the vegetation and climate under which they formed. Root traces, soil structure, and animal burrows reveal whether the vegetation was forest, sage scrub, or grassland. In modern soil of arid regions, nodules of calcite or dolomite are closer to the surface where annual rainfall is lower. The thickness of the soil layer containing such nodules indicates how seasonal this rainfall is. Other aspects of soil chemistry provide another estimate of mean annual rainfall, and a crude estimate of mean annual temperature. Retallack applied these rules to infer the climate of fossil soils.

Grassland evolution went through three successive stages, as more efficient grazers and browsers appeared. More than 30 million years ago, giant herbivores, *Titanotherium,* large rhinos, and primitive horses, *Mesohippus,* evolved teeth that could chew or shred tree leaves. These teeth enabled them to eat silica-rich grasses, although they preferred broad leaves of trees and shrubs. Eventually, huge herbivores evolved that, in scrub and woodland with 200–400 millimeters of rain a year, knocked down trees to eat their leaves, allowing bunch grasslands to replace the woody vegetation. By flooding these bunch grasses with light, they may have favored fast growth over costly defense.

About 20 million years ago, specialized grazers in Africa, Asia, and North America evolved while short sod grasslands (short-grass prairie) were replacing bunch grasslands. In short sod grassland, unlike bunch grassland, grass roots less than 40 centimeters deep form a continuous mat or sod, as in a lawn from which one rolls up strips of sod to replant elsewhere. A fossil short sod grassland soil at Fort Ternan, Kenya, closely resembles grassland soil from the modern Serengeti. It has fossil grasses, related to grasses in dry forests of Australia and Latin America, whose ancestors must have grown under dry forest or thorn scrub. Horses evolved high-crowned teeth to cope with the silica-rich grasses of short sod grasslands. These grasslands offered larger grazers no place to hide; they had to group like musk oxen to fend off predators, or outrun them. Horses, camels, and other grazers evolved the ability to run fast; now the English race horses, and Arabs race camels. Grazers that could run fast and far caused predators to evolve the same ability. New pack-hunting predators, dogs, forced grazers to live in herds, with more eyes to spot predators and more members to join in defense. The oldest known herd of grazers, a group of 17 camels, lived in Nebraska 19 million years ago. Herbivores evolved that dropped dung cakes, "cowpies," rather than rabbit pellets. Cowpies favored earthworms, more protein-rich grasses, and larger dung beetles. They fertilize grassland much better than the rabbit- or deer-like pellets of their ancestors, which are better fertilizer for forests. In response to predators, grazer herds grazed a patch of grassland down to the ground, tasty and unpalatable plants alike, plastering it with their dung, before moving on. Some recommend locking enough cattle into a fenced pasture for a day to graze and manure it thoroughly, then moving them on to repeat the process in the next pasture, as a good way to maintain pasture and grow cattle.

Finally, tall sod grassland, tall-grass prairie, evolved seven million years ago thanks to a new mechanism of photosynthesis—"C_4 photosynthesis," now used by corn plants and most tropical open-country grasses. This change enhances productivity in warm climates, for C_4 photosynthesis produces more carbohydrates per gram of water transpired. Tall grassland replaced dry forests with annual rainfall less than 700 millimeters in all continents except Australia, where sod grassland never evolved and "living fossil" mallee woodland grows where annual rainfall is 500 to 600 millimeters. As tall sod grassland evolved, herbivores evolved large hard hooves, which made it easier to run fast and to kick predators hard. Grassland was a new, productive habitat. Where grazers excluded trees, it offered livings for many new species, especially of mammals and birds.

The most famous grassland ecosystem is East Africa's Serengeti, with its magnificent herd of 1.5 million wildebeest, and many other grazers. Grazers consume 66 percent of the Serengeti's aboveground grass production, whereas vertebrate herbivores eat only 3 percent of Barro Colorado's leaf production. Grazing increases grassland production in the Serengeti by an average of 86 percent compared with ungrazed control plots, by renewing fresh nutritious new growth as long as the soil remains moist. Grazing reduces production of grass seeds by 90 percent but increases vegetative reproduction. Serengeti grassland needs elephants and other grazers to exclude trees, while the grazers depend on the grassland for a nutritious food supply whose abundance no other ecosystem can match (fig. 6.6). Grassland is the first ecosystem where folivores became mutualists of their food plants.

Fig. 6.6. In East Africa, grazing by mammals such as African elephants, *Loxodonta*, excludes trees from productive grassland. (Photograph by Christian Ziegler)

How Fragmentation Diminishes Competition, Diversity, and Interdependence

How thoroughly an ecosystem's function hinges on interdependence among different neighborhoods and regions becomes clear when a new reservoir "fragments" a forest into isolated islands. Damming Panama's Chagres River to form part of the Panama Canal created forested islands ranging from smaller than 1 to 1,500 hectares.

The largest island, 1,500-hectare Barro Colorado, suffered losses (fig. 6.7). This island was too small to support a herd of white-lipped peccaries, which normally have more than 100 animals; these peccaries disappeared in 1931 during a fruit famine. Its pygmy squirrels died out when very young forest grew up. Barro Colorado also lost night monkeys and four-eyed opossums. In 1930, sixteen years after isolation, this island had 200 species of resident breeding land birds, but by 1996, there were only 135. Thirty species of young forest birds died out as their forest aged. This island also lost 35 species of insect-eating birds of the forest understory,

Fig. 6.7. Fragmentation causes extinctions. Barro Colorado lost, among others, night monkeys, *Aotus*. (Photograph by Christian Ziegler)

including ocellated antbirds, dominant competitors for insects flushed by raiding swarms of army ants. Experiments showed that these insect eaters could not fly 100 meters over open water, so no immigrants replaced island populations that died out, or replenished genetic variation of surviving populations.

On the other hand, Barro Colorado's forest resembles intact mainland forest in most respects, because many animals can fly or swim between the island and the mainland. Many fruit-eating bats visit the surrounding mainland when they cannot find suitable fruit on the island. Before 2012, top predators, such as pumas and jaguars, visited the island to find prey. Immigrants swimming from the mainland have replenished genetic variation in ocelots, tapirs, tamanduas, monkeys, and the like. Some insect populations are restored by mainland immigrants after a food shortage or severe drought destroys them. Some migrating songbirds that breed in North America winter on this island. Some of Barro Colorado's bats and birds migrate to other tropical habitats to escape seasonal food shortage (plate 6.14). Parrots and toucans left Barro Colorado en masse during the fruit famine of 1969, returning after it ended. Preventing any of these movements would diminish the island's diversity.

Islands smaller than 1 hectare suffered far greater changes. They were too small to support any terrestrial mammals or to attract visitors from elsewhere. Losing these mammals disrupted some relationships of interdependence. Among those lost were agoutis, 8-pound rodents resembling the miniature forest antelopes of Africa, such as blue duikers (plate 6.15). Agoutis bury seeds, and dig them up again when food is scarce. Many tree species need agoutis to bury their seeds, which protects them from insect attack. On islets without agoutis, these tree species declined. Trees of species with large seeds, some of which could survive unburied—two species of *Protium,* two species of *Swartzia,* and the palms *Oenocarpus mapora* and *Attalea butyracea*—spread over these small islets, replacing trees with seeds that needed burial. Because fragmentation into islands disrupted many relationships between trees and their pollinators and seed dispersers, the diversity of trees on small islands plummeted. Seventy-six years after isolation from the mainland, more than half of the 400 trees with trunk diameter greater than 10 centimeters on each of two islets belonged to a single genus, *Protium.* No genus so dominates a 400-tree sample from Barro Colorado.

Herbivory is least on the smallest islands. On Barro Colorado, as in mainland forest, a species suffers most from pests where it is most common. Nonetheless, in the tree *Protium panamense,* caterpillars damage a significantly smaller proportion of leaves, and damage on injured leaves is less, on islets smaller than 1 hectare, nearly half of whose trees may be *Protium panamense,* than on Barro Colorado, where it is much rarer. Indeed, in each of four tree species shared by the 1,500-hectare Barro Colorado, 10-hectare islands, and 1-hectare islets—two species of *Protium,* and two understory species—the proportion of caterpillar-damaged leaves was lowest on the 1-hectare islets and highest on Barro Colorado, where these species were rarest. Moreover, injured leaves were least damaged on the smallest islets.

The pace of life is slower on small islands, as if competition is less intense there. Spiny rats, *Proechimys,* on 2-hectare islands live longer, and reproduce later and less often than mainland counterparts, which face many more predators. Islands no larger than 15 hectares have fewer insect-eating birds and more insect-eating lizards with far lower metabolic rates, because fewer predators eat lizards there. Even islands smaller than 1 hectare, however, harbor some energetic animals such as fig-eating and nectar-drinking bats, which roost there to escape predators, like seabirds on offshore islands.

What Factors Constrain Ecosystem Evolution?

When a new reservoir fragments a forest into islands, diversity declines, interdependence is disrupted, and the pace of life slows. Similarly, ecosystems of natural islands far more isolated than our fragments cannot evolve the diversity, productivity, intensity of competition, and levels of interdependence of mainland settings.

WHAT LIMITS DIVERSITY?

Réunion is a tropical island of 2,510 square kilometers, 800 kilometers east of Madagascar, created two million years ago by a great volcano rising above the ocean. Its highest mountain now rises 3,070 meters above the sea. Its habitats range from lowland savanna and rain forest to cloud forest and subalpine vegetation. Before Europeans arrived, however, Réunion had only 33 species of resident land birds; even now, 15-square-kilometer Barro Colorado has more than 100. Réunion has fewer than 660 native species of flowering plant; Barro Colorado, which is large enough, new enough, and near enough to the mainland to have lost few plant species since it was separated from the mainland in 1914, has more than 1,000. Moreover, 1,000 contiguous trees from everwet lowland forest of Réunion with a trunk diameter of at least 10 centimeters include only 40 species, compared with more than 120 among 1,000 trees from seasonal forest on Barro Colorado and more than 300 among 1,000 trees from everwet forest in Sarawak, Malaysia.

WHAT LIMITS THE INTENSITY OF COMPETITION AND THE PACE OF LIFE?

Competition is weakest on small, isolated islands. Rain forest on the 10,000-square-kilometer "Big Island" of Hawaii is half as productive as continental rain forest on poor soil, and its wood production—an index of competition for light—is especially low. Herbivores and carnivores are smaller on smaller islands, so herbivore and predator pressure, major components of competition, are lower there. Two physiologists, Gary Burness and Jared Diamond of the University of California at Los Angeles, and an Australian mammalogist, Tim Flannery, tallied the weight of the largest herbivore and carnivore on land masses of different size (fig. 6.8). Where people settled only recently—Australia, the Americas, Madagascar, and the Pacific islands—fossils were used to estimate the weight of the largest herbivore and carnivore present before settlement. In both, the weight of a land mass's largest species was roughly proportional to the square root of its area. Before people arrived, the largest herbivore and carnivore on Mauritius, a 1,900-square-kilometer oceanic island 1,000 kilometers east of Madagascar, were the 19-kilogram dodo and a 600-gram hawk, whereas on the long-isolated tropical island of Madagascar, 587,000 square kilometers (bigger than France), they were a 400-kilogram pygmy hippopotamus and a 17-kilogram *Cryptoprocta,* a catlike member of the mongoose family. Africa, however, was big enough to allow 5,500-kilogram elephants and 200-kilogram lions to evolve.

Presumably because herbivore pressure is so low, Hawaii has almost no native poisonous plants. As the naturalist Sherwin Carlquist observed, Hawaiian mints have lost their minty flavor and Hawaiian nettles their sting. Some Hawaiian plants evolved prickles to deter giant leaf-eating ducks and geese, which did not protect them from introduced pigs. Hawaii's forest suffered grievously from introduced mammal herbivores, but its forest suffered little from introduced insects. Despite long exposure to giant herbivorous moas, even New Zealand's plants are far more vulnerable to vertebrate herbivores than mainland counterparts. Madagascar, however, hosts a spectacular "arms race" between bamboos loaded with cyanide and

Fig. 6.8. Larger land masses support larger herbivores: A 5-ton forest elephant, *Loxodonta,* 3 meters (10 feet) tall, from Africa, compared with a 19-kilogram (40 pound) dodo, *Raphus,* 90 centimeters (3 feet) tall, which before human beings settled was the largest herbivore on the oceanic island of Mauritius. (Drawing by Damond Kyllo)

cyanide-tolerant bamboo-eating lemurs (fig. 6.9). Moreover, lowland grassland evolved on no island smaller than Madagascar, although many grasses reached Hawaii. Sod grassland never evolved in Australia, only bunch grassland. For grasslands to evolve, herbivore pressure must be intense. In Madagascar, grazing hippopotami that immigrated from Africa probably caused grassland to evolve.

Reduced predation affects island life in many ways. Animals on isolated islands live longer and reproduce more slowly. In Hawaii, native tree snails, *Achatinella* and *Partulina,* have no young until they are six or seven years old, and produce only six or eight young a year thereafter. In contrast, an introduced snail that eats them, *Euglandina rosea,* lays 600 eggs before the end of its first year. New Zealand, which separated from Australia and

Fig. 6.9. Unlike the Hawaiian Islands, where herbivore pressure is low, even Madagascar has intense arms races between plants using poisons to make their leaves less appealing to leaf eaters and herbivores eating these leaves. Here a cyanide-resistant lemur, *Prolemur simus,* which eats bamboo shoots, *Cathariostachys,* laced with cyanide. (Photograph by Christian Ziegler)

Antarctica more than 65 million years ago, is notorious for its long-lived, slowly reproducing animals. Its endemic brown geckoes, *Hoplodactylus,* are the longest-lived, slowest-reproducing lizards known. Female tuataras, *Sphenodon,* New Zealand's famous "living fossil" reptiles, whose adults weigh 1.2 kilograms, first reproduce when thirteen years old, and produce fewer than 20 eggs every four years. In contrast, female green iguanas of Barro Colorado, *Iguana,* whose adults weigh less than 3 kilograms, first lay eggs when two to eight years old, and lay an average of 41 eggs every year or two. Longer life and slower reproduction has other consequences. Brian McNab, a physiologist at the University of Florida, found that the metabolic rates of pigeons, fruit-eating bats, and rails on South Sea islands of 1,000 square kilometers are markedly lower than those of ecological counterparts on land masses exceeding 100,000 square kilometers.

INTERDEPENDENCE IS WEAKER ON SMALL, ISOLATED NATURAL ISLANDS

Madagascar is famous for *Angraecum* orchids, which are pollinated only by moths with a very long proboscis (tube) through which they suck nectar. An *Angraecum* flower chooses its pollinators by keeping its nectar in a "spur," a deep narrow container just right for the proboscis of its favored moth to reach its bottom. The flower deposits pollen on a specific part of the moth, which will touch the stigma (pollen receptor) of the next flower it visits. Darwin was shown flowers of the Malagasy orchid *Angraecum sesquipedale,* with nectar-holding spurs 30 centimeters deep. He predicted the discovery of a sphinx moth with a proboscis long enough to reach this nectar. Such a moth, *Xanthopan morganii praedicta,* was found in 1903, but was not shown to pollinate *Angraecum sesquipedale* until 1997. On Madagascar, many orchids attract specialized pollinators.

Several different Malagasy species of *Angraecum* orchids colonized Réunion, a far smaller island 800 kilometers from Madagascar. Their descendants responded to the less competitive conditions on Réunion in various ways. All the *Angraecum* on Réunion whose flowers have spurs over 9 centimeters long, originally meant to hold nectar to attract pollinating moths, now pollinate themselves without help from any animal. Very few orchids of the *Angraecum* subtribe in Madagascar, but a third of those endemic to Réunion, are self-pollinated. One *Angraecum* from Madagascar, which may have reached Réunion before sphinx moths did, now has three descendant species there (fig. 6.10). The scentless flowers of one, *Angraecum striatum,* whose wide-mouthed nectar spurs are only 1.2 centimeters deep, are pollinated only by an abundant small native bird, the white-eye *Zosterops borbonica.* This bird eats fruits, seeds, and insects as well as nectar, whereas the moths attracted by the heavily scented flowers of this orchid's Malagasy ancestors were specialized orchid-pollinators. Another, *Angraecum bracteosum,* whose equally scentless flowers have spurs only 0.8 centimeter deep, is pollinated mainly by *Zosterops olivaceus,* which collects nectar from many kinds of flowers. The third, *Angraecum cadetii,* is pollinated only by a wingless "raspy cricket," *Glomeremus* sp. Its flowers have

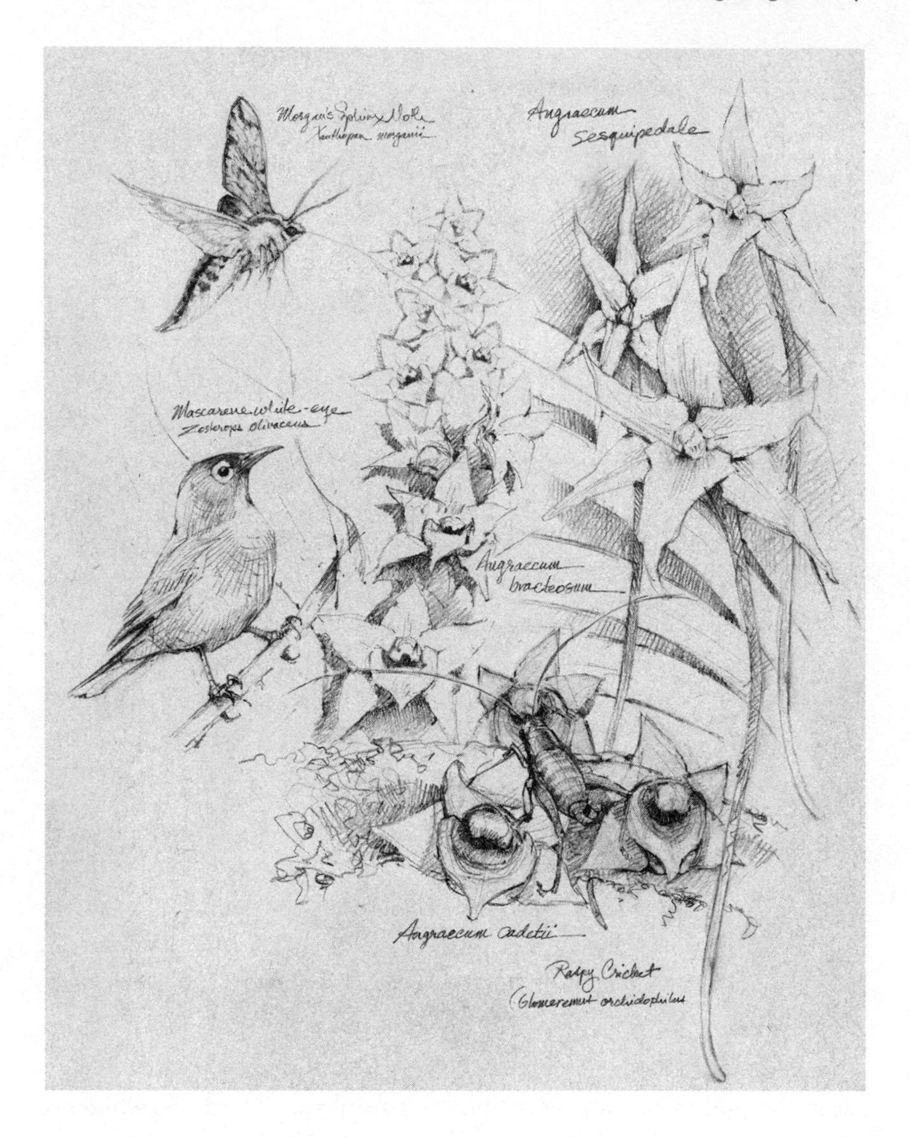

nectar spurs only 0.6 centimeter deep. Its pollinating cricket eats seeds, pollen, other vegetable matter, and a few arthropods as well as nectar. Since it cannot fly, it must walk. This pollinator is much less effective, but much cheaper, than the demanding but far-flying sphinx moths that pollinated this orchid's ancestors in Madagascar.

Indeed, lower diversity on small islands both reflects and causes reduced opportunities for interdependence, while weaker competition on these islands reduces the need for help from other species to compete effectively. On small islands, insects are relatively rare and herbivore pressure low. Plants of a particular species need not be sparsely scattered to escape insect attack, so they do not need pollinators willing to travel far to find another plant of their species. Indeed, on small oceanic islands many insect pollinators are super-generalists that serve many plant species. Native plants are more likely than mainland counterparts to have small, dull-colored, odorless flowers. Few plants of island species compete vigorously for the insect pollinators available. As we have seen, some employ unspecialized birds or flightless insects. Others are wind-pollinated. Because they need less genetic variability to cope with pests and disease, island plants depend far less on outcrossing—being fertilized by pollen from other plants—than mainland counterparts, and they lack features that prevent them from "selfing," being fertilized by their own pollen. Indeed, island plants have abandoned the evolutionary achievements of the flowering plants' last hundred million years, by which they coped with pests and disease.

HOW ISOLATION WEAKENS COMPETITION AND MAKES IT EASIER FOR EXOTICS TO INVADE

Darwin observed that lowlands of small isolated islands were populated mainly by weedy species that people introduced from the mainland. He

---

Fig. 6.10. (*opposite*) On large land masses, orchids use high-performance pollinators, whereas on small oceanic islands, where specialist pests do not keep their food plants so rare, orchids make do with far less specialized pollinators that move shorter distances. At far right is a long-spurred orchid, *Angraecum sesquipedale,* from Madagascar, being pollinated by a sphinx moth at top left with a long proboscis, *Xanthopan morganii praedicta,* whose existence Darwin predicted from the orchid's long spur. At center is an orchid of the same genus from Réunion, *Angraecum bracteosum,* whose pollinator is the generalized nectar-eating Mascarene white-eye, *Zosterops olivaceus,* at center left. At bottom is another Réunion orchid, *Angraecum cadetii,* whose pollinator, the raspy cricket *Glomeremus orchidophilus,* cannot even fly. (Drawing by Debby Cotter Kaspari) argued that the lower intensity of competition on oceanic islands, especially small ones, left them devastatingly vulnerable to introduced animals and plants, whose mainland ancestors had survived interactions with many competitors and predators. In New Caledonia, where ants are moderately diverse, an ant accidentally imported from South America, the little fire ant, is replacing all of the native ants it encounters. These fire ants find food more quickly than native ants, and summon other workers quickly enough to defend their finds against them. In isolated oceanic islands such as the Hawaiian Islands, Réunion, and Mauritius, invasive plants exploit light and nutrients more effectively than native counterparts. The ineptitude of native predators makes their islands more vulnerable to invasion and disruption by introduced species. The slow reproducers on isolated islands are quickly ousted by active, adaptable, voracious, fast-reproducing animals introduced from the mainland. Rats from Eurasia spread over New Zealand's main islands in a century or two, eliminating slow reproducers such as the largest native frogs and skinks, tuataras, and large long-lived snails and orthopterans, or confining them to offshore islands rats never reached. In Madagascar, a larger, less isolated island with effective carnivores descended from mongeese that crossed from Africa only 20 million years ago, introduced rats spread far more slowly and caused far fewer extinctions. Introduced rats cannot invade undisturbed Neotropical rain forest: there, they survive only as commensals in human settlements. When a graduate student, Robert Kimsey, fed up with the rats teeming in Barro Colorado's laboratory clearing, trapped them all, there were none to recolonize from the surrounding forest.

Interdependence among invaders enhances their success. Introduced animals often speed the spread of introduced plants. In the Hawaiian Islands, introduced birds and pigs eat the tasty fruits of strawberry guava, an introduced shrub, and disperse its seeds. In Mauritius, introduced pigs and deer clear openings for strawberry guava and many other invasive plants. The Black River Gorges National Park in Mauritius is flanked by thickets of guava seemingly too dense for a dog to slip between the stems (plate 6.16). A few surviving native trees tower over this thicket, but the forest floor is now too dark for their seedlings to survive. In Réunion, introduced mammals have been excluded or their abundance limited, so much more native forest remains.

Long isolation made even continents as large as Australia and South America vulnerable to invaders from larger, less isolated continents with more intense competition. When placental predators, such as dogs and big cats, crossed the new isthmus of Panama from North to South America, they quickly replaced South America's marsupial carnivores. Likewise, wild dogs, dingos, invading Australia 6,000 years ago from southeast Asia quickly replaced slower-moving, less effective marsupial carnivores—Tasmanian wolves, *Thylacinus,* and Tasmanian devils, *Sarcophilus.* These marsupials survived in nearby Tasmania, an island that dingos never reached. Marsupial predators could not compete with placental counterparts because their design limited their metabolic rates and activity levels. Isolation, however, protected them from placental competition. When placental predators arrived, larger grassland mammals, totally unprepared for these agile carnivores, were devastated. Even occasional immigrants, however, reduce the impact of later invaders. Thanks to the placental mongeese that invaded Madagascar 20 million years ago, introduced cats, rats, and dogs damaged Madagascar's fauna far less than larger but more isolated Australia's. Natural economies must be large, and at least intermittently connected to others, to develop the intensity of competition and relationships of interdependence that underlie the high diversity and productivity and competitiveness needed to resist exotic invaders.

This chapter showed how the interplay among adaptation of individuals, social cooperation, diversification of species, competition for resources, interdependence within and among regions, and cooperation between members of different species brought forth productive, diverse natural communities, resistant to disruptive invasion. These processes are more effective on larger land masses. In natural ecosystems, as in human economies, self-sufficiency is the enemy of productivity and diversity. What drove these developments? To answer, we show how natural selection works, and how it shaped genetic systems allowing complex organisms like trees and human beings to evolve.

SEVEN

Heredity, Natural Selection, and Evolution

DARWIN PROPOSED THAT NATURAL selection, whereby individuals better suited to survive and reproduce have more offspring that survive to reproduce in their own turn, is what drives adaptation. To put it crudely, natural selection occurs because better individuals have more and better offspring. Therefore, as Darwin understood as well as anyone, understanding how selection influences evolution requires knowing how offspring inherit their parents' characteristics.

Natural selection, however, does not always promote an individual's own reproduction. A leaf-cutter or army ant colony contains many sterile workers but only one reproductive queen: how can selection favor this circumstance? A first hint comes from fairy tales, which often reflect familiar human experience. A central theme of such tales is the cruel stepmother, the father's second wife. Many second wives are, of course, quite kind to their predecessor's children, but it is close to a law of nature that if they play favorites, they favor their own children over their predecessor's. Today, we hear more about the new boyfriend, who is far more likely to mistreat his predecessors' children than his own. Indeed, preferring the welfare of relatives at the expense of non-relatives happens so often, in so many contexts, that there is a special word for it, "nepotism." The mechanics of heredity help to reveal why nepotism is so prevalent.

Other animals help relatives. Leaf-cutter ant workers, who are sterile, instinctively feed their mother, the queen ant, and raise her young. Honeybee workers help their mother, the queen bee, reproduce, as long as she keeps doing so. How this assistance is enforced is one of the great wonders of nature. A honeybee colony functions because workers help closer relatives even if this means killing more distant ones. Workers can lay male eggs without mating, and some do so rather than help their queen. The queen suppresses such "cheating" by pairing with many drones on her one mating flight, and mixing their sperm, making most of a worker's fellows half sisters with different fathers. Bee larvae must be cared for, and the care available is limited. A worker is more closely related to her queen's eggs than to her half sisters', so she eats other workers' eggs, creating an effective common interest among workers in helping their queen reproduce rather than laying eggs of their own.

Why should animals help relatives? What does the tendency to favor closer relatives tell us about what natural selection promotes? The secret lies in how heredity works. So, what factors govern our resemblance to, and our heritable differences from, our parents? Characteristics imposed on a plant or animal after its birth are not inherited. Cutting off a dog's tail does not cause its offspring to lack tails. Indeed, what is inherited is not a "blueprint" of the finished plant or animal but a recipe, a set of instructions on how to make it, starting from a single fertilized egg—that programs an ordered series of processes, as a computer program specifies an ordered sequence of operations. Given the right setting, resources, and stimuli, this recipe usually causes a fertilized egg to grow into an adult of its parents' kind. How are these instructions passed from parent to offspring?

How Mendel Deduced and Proved the Laws of Genetic Inheritance

For sexually reproducing plants and animals, a Czech monk, Gregor Mendel, guessed that: 1, the instructions for each basic process are embodied in particular factors now called the genes for that process; 2, an individual has two genes for each process, one from each parent; 3, when "gametes"—sperm in males, eggs in females—are formed, each gamete receives only one gene for each process, chosen at random from the two in the parent; 4, once fertilized, the egg, the new individual's founding cell, has two genes for each process, one from each parent.

Mendel tested his guesses by experiments with pea plants. His experiments required alternative versions of the instructions for the same basic process. In other words, he needed two kinds of genes, two *alleles,* for the same process, each allele carrying different instructions that program contrasting effects on the same visible characteristic. In pea plants, he supposed that one allele governing the seed's surface programs round seeds, and the other, wrinkled seeds, and that one seed-color allele programs yellow seeds, and the other, green seeds. To test his ideas, he first crossed pea plants of a strain with round, yellow seeds, presumed to have two round-seed and two yellow-seed genes apiece, with plants of another strain, presumed to have two wrinkled-seed and two green-seed genes apiece. These produced hybrid young. He predicted that each hybrid seed should have one round-seed gene and one wrinkled-seed gene, *and* one yellow-seed and one green-seed gene. Yet all hybrid seeds were round and yellow. Does this mean that in a seed with one round-seed gene and one wrinkled-seed gene, the round-seed gene "dominates" the wrinkled-seed gene, making the seed round? Similarly, does the yellow-seed gene dominate its green-seed counterpart?

To find out, Mendel "selfed" these hybrids, fertilizing the flowers of each hybrid with its own pollen. If half of the gametes of each hybrid—half its pollen grains and half its ovules or egg cells—carry a round-seed gene, and half, a wrinkled-seed gene, then half the pollen grains with round-seed genes will fertilize ovules with round-seed genes, and half, ovules with wrinkled-seed genes (fig. 7.1). The same will be true for wrinkled-seed pollen grains. Thus a quarter of these selfed hybrids' seeds should carry two round-seed alleles, another quarter, two wrinkled-seed alleles, and half, one of each. If "heterozygous" seeds with one round-seed and one wrinkled-seed allele are all round, like the seeds that gave rise to their hybrid parents, three fourths (75 percent) of these hybrids' seeds will be round, and one fourth (25 percent), wrinkled. Of 556 seeds Mendel obtained by selfing hybrids, 423 (76.1 percent) were round, and 133 (23.9 percent), wrinkled. Similarly, 416 (74.8 percent) of these 556 seeds were yellow, and 140 (25.2 percent), green. He also found that two thirds of the yellow seeds of selfed hybrids produced heterozygous plants which, when selfed, produced some

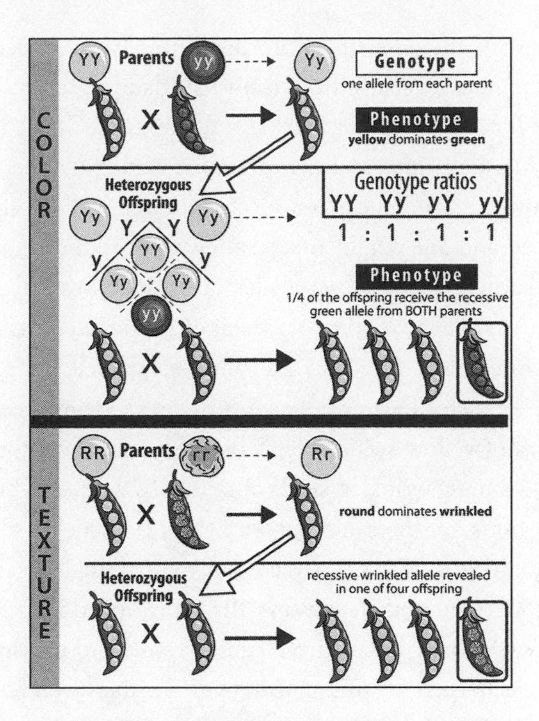

Fig. 7.1. Mendel's first deduction: an organism inherits one gene at each locus from each parent. Upper panel: color locus in peas. YY homozygotes and Yy heterozygotes produce only yellow peas, and yy homozygotes only produce green peas. If one crosses two Yy plants, half the pollen gametes carry Y, half y, and each gamete is equally likely to fertilize an ovule with a Y or a y, so in a quarter of the offspring, two Y gametes unite, creating YY genotypes with only yellow peas. Similarly, a quarter will be yy, with only green peas, while the remainder are Yy heterozygotes with only yellow peas. Lower panel: seed surface locus in peas. RR homozygotes and Rr heterozygotes produce only round peas, and rr homozygotes, only wrinkled peas. If one crosses Rr with Rr, a fourth of the offspring will be rr homozygotes with wrinkled peas. (Diagram by Damond Kyllo)

green seeds, while the other third produced only yellow seeds. Similarly, two thirds of the plants grown from round seeds of selfed round-wrinkled hybrids were heterozygotes for seed surface texture. These findings confirmed both Mendel's idea and the existence of dominance, where seeds with one yellow-seed and one green-seed allele were yellow, and those with one round-seed and one wrinkled-seed allele were round.

Let an individual have two alleles at both the round-seed and the yellow-seed locus. Then, at both loci, each of its gametes has equal chance of receiving either allele. If receiving a round-seed allele does not affect the chance that a "gamete" (egg or pollen grain) gets a yellow-seed allele, then the proportion of yellow seeds should be the same, three fourths, among round seeds as among wrinkled seeds obtained by selfing hybrids (fig. 7.2). These proportions are the same: 315/423 (74.5 percent) of these hybrids' round seeds, and 101/133 (75.9 percent) of their wrinkled seeds were yellow. Thus, as Mendel predicted, a sexually reproduced individual receives one gene for each basic process from each parent, and transmits one gene for each process to each offspring. Moreover, for the seven basic processes Mendel published on, inheriting a particular allele for one process does not affect the probability of inheriting a particular allele for a different process. Genes for these different processes are inherited independently, as round-seed genes were inherited independently of yellow-seed genes.

Mendel's paper came out in 1866, but it lay ignored until 1900, when studies of chromosomes suggested that they were the carriers of heredity. Chromosomes are elongate objects in the nucleus of eukaryote cells, which come in pairs. As a rule, all of the cells belonging to the same species, gametes excepted, have the same number of pairs of chromosomes. Just before a cell divides in the course of ordinary growth, each of its chromosomes is copied. The process of cell division seems designed to ensure that each daughter cell gets one copy of each chromosome. When gametes are produced, two successive cell divisions cause each gamete to be randomly assigned one chromosome of each pair. In short, these chromosomes behave as if each one carried one of Mendel's genes, where the genes on one pair of chromosomes program the same basic processes, whereas genes on other pairs program different processes. Further work showed that each

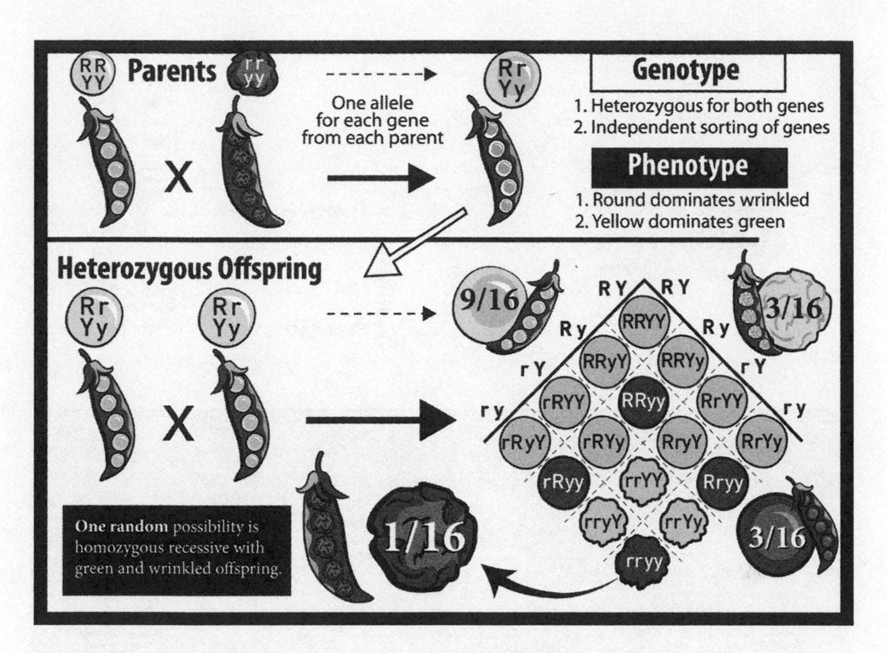

Fig. 7.2. Mendel's second deduction: independent inheritance of genes at different loci. Mendel predicted that if we cross RRYY homozygotes making round yellow peas with rryy homozygotes making green wrinkled peas, we will get RrYy genotypes, heterozygous at both loci, making round yellow peas. Then, if one trait does not affect the other and assorts independently of the other, ¼ of the pollen grains will carry RY, ¼ Ry, ¼ rY, and ¼ ry, and each gamete will have an equal chance of fertilizing an RY, Ry, rY, or ry ovule. Thus, if we mate two RrYy plants, 9/16 of the offspring should have round yellow peas, 3/16 wrinkled yellow peas, 3/16 round green peas, and 1/16 wrinkled green peas. Mendel's experiment fulfilled this prediction. (Artwork by Damond Kyllo)

chromosome carries a linear array of genes arranged in a fixed order. Not only do chromosomes of a pair carry genes programming the same set of basic processes, these genes are arranged in the same order. Thus a gene for a particular process, for example, the gene coding for amylase, an enzyme in our saliva that breaks down starch, a chain of sugars, into its component sugars, occupies a particular spot, called the *locus* for that process, on its chromosome (fig. 7.3).

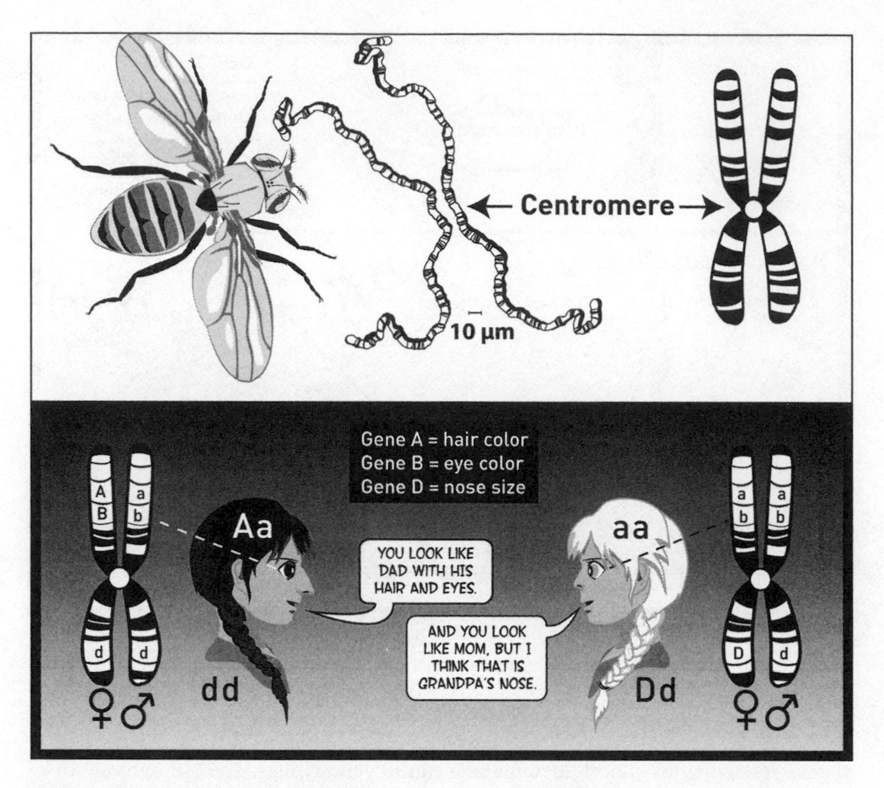

Fig. 7.3. Where genes reside: Diagram showing a fruit fly, *Drosophila melanogaster,* one homologous pair of its chromosomes, and a schematic cartoon of this pair. A chromosome embodies a sequence of loci governing processes enabling the organism's function. Different species have different numbers of chromosomes; human beings have 23 pairs of chromosomes, *Drosophila,* four. In each pair of homologous chromosomes, one chromosome is from the mother, one from the father. Different genes affect different characters. The bottom panel shows how, by virtue of receiving different chromosomes from their parents, two sisters have genotypes programming different hair color, eye color, and nose length. A daughter can inherit either of two homologous chromosomes from her mother, and likewise from her father, which is why the sisters' genotypes can differ. (Artwork by Damond Kyllo)

How Organisms Read Their Genes' Instructions

How does a gene store the instructions for its process? James Watson and Francis Crick showed in 1953 that its instructions are encoded in its sequence of "nucleotides"—adenine, thymine, guanine, cytosine—as Morse code encodes a message in a sequence of dots and dashes. In parts ("exons") of a functional gene, each triplet of nucleotides specifies a particular amino acid; thus a codon of three adenines codes for the amino acid lysine. The sequence of triplets programs a sequence of amino acids that are strung together to make a protein, which then folds automatically into the shape that makes it do its job (fig. 7.4). This protein is often an enzyme that catalyzes a particular biochemical reaction important to its bearers. Other parts of this gene are sites where different repressor proteins can bind to different sequences of codons, allowing this gene to act only when needed. For example, one gene's enzyme catalyzes the split of lactose, a milk sugar, into more usable sugars. A repressor of this gene normally blocks this enzyme's synthesis. When lactose is present, however, the repressor combines with it, so this enzyme is synthesized only when lactose is available to split. The repressor of a gene whose enzyme is involved in making a particular amino acid may block that enzyme's synthesis only when combined with that amino acid, so the enzyme is not synthesized when the bacterium can ingest that amino acid from its surroundings. Other repressors act so as to coordinate the activities of different genes.

A catalyzed reaction can have many effects. In butterflies, the first step in breaking down glucose to supply fuel to the flight muscles is a reaction catalyzed by the enzyme phosphoglucose isomerase (PGI). A sulphur butterfly's PGI genes influence how long it can fly, how early it can get off the ground, the range of body temperatures at which it can fly, and how well it survives hot weather. Being able to fly for long periods enables males to court females more successfully, females to find better places to lay their eggs, and both sexes to find the nectar they need to live, and to deal with sudden changes in weather. Both sexes escape predators better when they can fly vigorously. In sum, the reaction catalyzed by this one enzyme, PGI, matters greatly to these butterflies' ability to survive and multiply.

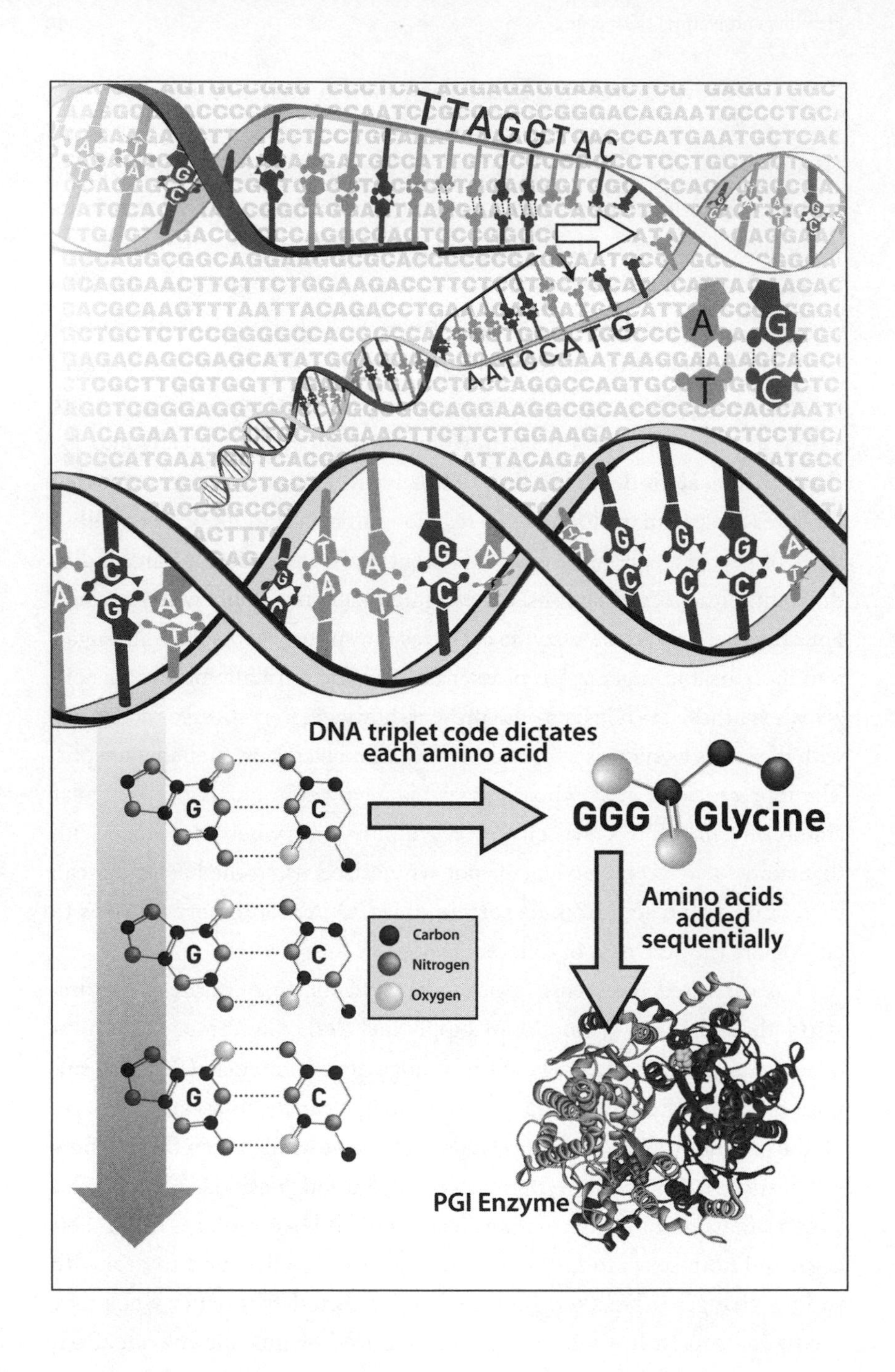
TTAGGTAC
AATCCATG
DNA triplet code dictates each amino acid
GGG
Glycine
Carbon
Nitrogen
Oxygen
Amino acids added sequentially
PGI Enzyme

Biologists have long wondered how a genome could program differences among an animal's cells when all had the same genome. Fruit flies, *Drosophila,* provided a schematic answer. In each of its eggs, the mother programs a gradient declining from the egg's front to its rear in the concentration of one chemical, one declining from rear to front in another, and one declining from bottom to top in a third. These gradients persist after the egg is partitioned into cells. Certain regulator genes are expressed only at specific levels of each gradient, so different combinations of genes will be expressed in parts of the embryo with different levels of each gradient. Each combination of expressed genes allows only those reactions, especially those development-related reactions, appropriate to its part of the embryo. In particular, they determine where crucial switch genes, such as those promoting eye, heart, and limb development, can act. Switch genes also act by turning others on and off. Many are shared by all bilaterians. If one replaces a fruit fly's normal allele at the eyeless locus, which promotes eye development in fruit flies, by the normal allele at the small-eye locus, which promotes eye development in mice, the mouse gene promotes the growth of compound fruit fly eyes, not single-chamber mouse eyes. Thanks to those gradients, location regulators, and switch genes, one genome can give rise to an organism with many different tissues and organs.

We cannot yet tell what basic process a gene programs by reading its

---

Fig. 7.4. (*opposite*) Schematic illustration of replicating DNA. The strands of DNA helix have separated, and each is acquiring a complementary strand. Each base can attach to only one other base, different from itself, so one DNA strand's base sequence dictates the other's. In fact, DNA replication must be catalyzed by complex enzymes. Below, how DNA codes for enzymes, showing two codons (sequences of three base pairs, each sequence corresponding to an amino acid), with a detailed diagram of the codon for the amino acid glycine. The code consists of four "letters," the bases adenine (A), thymine (T), guanine (G), and cytosine (C). The gene's message, encoded as a sequence of bases, is translated into a sequence of amino acids that folds up automatically to make an enzyme: phosphoglucose isomerase (bottom right), which catalyzes a reaction involved in the release of energy to flight muscles. This enzyme is composed of two chains of 555 amino acids each. (Artwork by Damond Kyllo)

DNA as one learns how to assemble a toy by reading an instruction manual. We are like the being the French novelist Georges Bernanos imagined "with an intelligence completely different from ours, thoroughly ignorant of language and writing, and totally illiterate, (who) might go into ecstasies over . . . the symmetry of a page of printing . . . without really knowing anything about . . . what is really important about all this, the thought, the thought always alive and free under the apparent constraint of the characters . . . that express it."

Although we cannot "read" DNA in any meaningful sense, Watson and Crick's discovery had profound effects. To begin with, in all forms of life, genes use the same code to program their basic processes, thus testifying to their common ancestry. We also construct the "family tree" of a set of species by comparing the DNA sequences of genes that program the same process in different organisms. Genes, too, descend with occasional modifications—mutations, copying errors. The more distant the ancestor, the more modifications in a gene inherited from that ancestor. Therefore genes for a process differ more between less closely related species (fig. 7.5). Species whose genes for a process have more similar DNA sequences usually have more recent common ancestors. As was mentioned earlier, Aristotle concluded that whales were more closely related to land mammals than to fish: the fossil record confirms that ancestral whales diverged from land mammals much more recently than ancestors of mammals diverged from fish. Moreover, corresponding genes from whales and human beings are much more similar in DNA sequence than either are to their counterparts in any fish. Where the fossils are preserved, the family tree inferred from these fossils usually matches that inferred from resemblances among genes programming the same basic process.

In sum, as a metazoan's cells multiply, each cell gets the same two sets of genes as its parent, barring the occasional miscopied gene, a "somatic mutant." Before gametes are produced, each pair of chromosomes exchanges corresponding parts, with genes for the same operations, in a process called recombination. Each gamete, however, gets only one chromosome of each pair, and one gene for each basic operation, as Mendel foresaw.

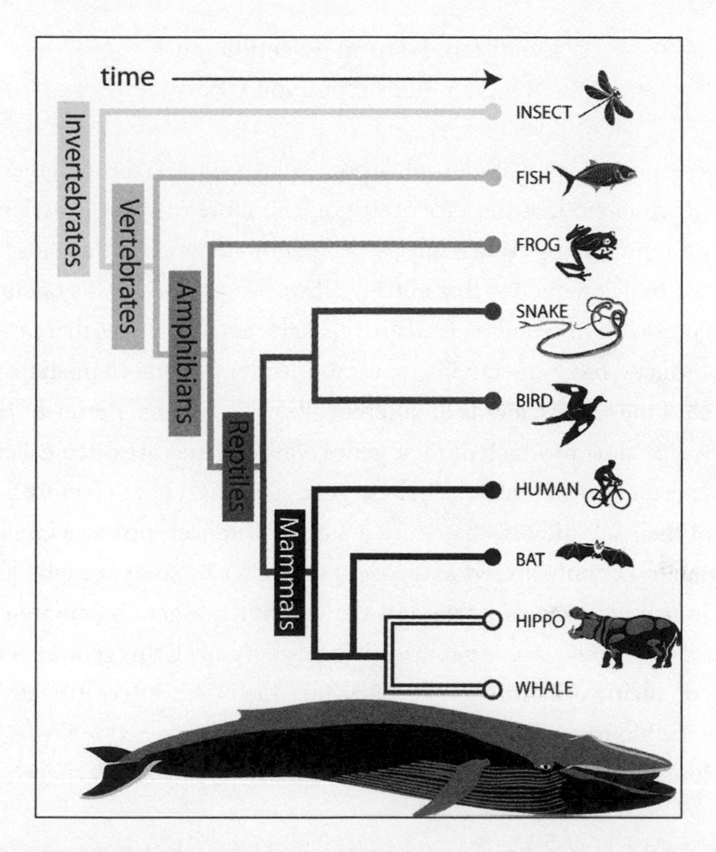

Fig. 7.5. The longer ago two groups divided, the more opportunity for genetic change in each, and the greater the difference in DNA sequence of a standard sample of genes they share. Shown is a phylogenetic tree representing the order in which the different groups shown diverged. This order is the same whether inferred from the divergence in DNA sequence in the genes sampled from each pair of groups, or comparative anatomy, or the fossil record. Whales group with hippos because their DNA sequences are most similar. The fact that whales and fish both swim in the ocean has no more bearing on their relatedness than does the fact that bats and insects both fly. On the other hand, the fact that whales and people have lungs and warm blood, and give birth to live young that had been nourished by a placenta, reflects that they have more recent common ancestry. (Artwork by Damond Kyllo)

Evolution, Natural Selection, and the Genome's Common Good

A species undergoes evolution when some alleles spread, through their carriers' reproduction, at the expense of others, more of whose carriers die without reproducing. When alleles become more common because they program basic operations that improve their bearers' prospects of survival and reproduction, evolution is adaptive. Genes embody codes that program or coordinate basic operations; natural selection is a mechanistic consequence of the relative merits of different alleles' programs. Because species evolve according to which of their genes spread, genes are often called the "ultimate units of self-interest." These genes, however, have no more awareness of their self-interest than does a stone. Moreover, just as a computer program works only if used in the right computer by someone who knows how to use both the program and the computer, a gene's program only works if its genome is in the appropriate setting, and all this genome's other genes do their work. Thus these "ultimate units of self-interest" really represent the ultimate in interdependence. What natural selection favors must therefore be the common interest of an organism's genes. How can we show this?

Nearly always, the process governing which of the two genes at a heterozygous locus a gamete receives is one of the fairest lotteries known. If natural selection spreads "cheater" alleles that bias this process, making most of a heterozygote's gametes carry the cheater allele, evolution would not necessarily be adaptive. Like cheating on an examination, a cheater allele's biasing meiosis in its own favor gives the cheaters an "unfair" advantage over their competitors. A few species have a locus where a "cheater" allele spreads by getting into "more than its share" of the gametes. If, however, meiosis was biased at most loci, an allele's spread would reflect its contribution to its bearers' fitness as poorly as the test score of someone who cheated at the examination reflects his knowledge of the material. The fairness of the lottery assigning genes to gametes serves the common interest of an individual's genes by ensuring that an allele spreads only by benefiting its bearers, and/or their relatives, enhancing their survival and reproduction. In particular, the fairness of this lottery ensures that selection coordinates the activities of different genes so as to make their bearers functional.

The common interest of an individual's genes is often expressed by favoring blood relatives, which share a disproportionate number of their bearer's alleles. Thus an individual serves its genes' common good by helping close relatives reproduce—when this does not place its own survival or reproduction at excessive risk. Renaissance popes promoting their nephews, honeybee workers caring for closer, at the expense of more distantly, related eggs, and many other phenomena involve helping close relatives.

In sum, instructions for producing the characteristics parents transmit to offspring are encoded in genes. A gene occupies a particular locus or "address" on a chromosome, and programs an operation peculiar to that locus, influencing some aspect of an individual's growth or behavior. Genes at the same locus programming different versions of its operation are said to represent different alleles. In animals, an individual inherits two genes at each locus, one from each parent. Thanks to the rigid fairness of the rules by which genes are passed from parent to offspring, natural selection on genes drives the adaptation of individuals and the diversification of species. To learn how the spread of genes adapts individuals to their ways of life, read on.

EIGHT

Organizing Genes for Adaptive Evolution

TO SHOW HOW EVOLUTION WORKS, WE must explain two things. First, the most obvious feature of living things is adaptation, their aptness for procuring what they need to live and multiply. How can natural selection on "selfish" genes be made to yield adapted individuals? Second, how did life's many forms diversify from microscopic common ancestors?

Natural Selection: Some Examples

First, let us consider simple examples where natural selection of specific alleles produced better-adapted individuals. Natural selection occurs when one allele's version of a basic program improves its bearers' capacity to multiply, enabling this allele to spread and replace its alternatives. Thus natural selection on mosquitoes spreads alleles arising by copying error (mutation) that happen to improve their resistance to an insecticide used against them. Resistant mosquitoes—and the alleles causing their resistance—survive to multiply; non-resistant counterparts do not. These mosquitoes thus become progressively more resistant to the insecticide in question. Similarly, in bacteria causing diseases such as tuberculosis and pneumonia, alleles spread that enhance resistance to the drugs used to cure them. In all these cases, a miscopied allele spread because it allowed a vital basic process, formerly blocked by the insecticide or drug, to proceed in its presence. Often, one can identify the succession of mutations—copy errors—creating new alleles whose DNA encodes improved resistance to a new drug. For example, some bacteria evolved an enzyme, β-lactamase,

Fig. 8.1. Predation by birds favors animals that match their backgrounds: Dark and light forms of the peppered moth, *Biston betularia,* on a dark background. (Photograph by Stephen Dalton/Minden Pictures)

that breaks down penicillin and related drugs, "β-lactams," into components that are no longer poisonous. A new β-lactam, cefotaxime, killed these bacteria because their β-lactamase broke this drug down too slowly. Five nucleotide changes (five single-letter changes in the DNA code), however, arose in genes for β-lactamase, each enabling their bearers to multiply faster than their predecessors by increasing their resistance to cefotaxime. Collectively, these five mutations increased their bearers' resistance to cefotaxime 100,000-fold. Here, each stage in the evolution of resistance to the drug has been identified—both the change in the DNA code causing it, and its contribution to cefotaxime resistance.

Other environmental changes favor visible evolutionary change. In Britain, "peppered moths," *Biston betularia,* normally rest during the day on the undersides of lichen-covered tree branches: its speckled color closely matches these backgrounds (fig. 8.1). In the 1800s, factories produced sulphur dioxide and soot, killing lichens and blackening trunks and branches in nearby forests. Moths with an allele that darkened body and wings, making its bearers match sooty, blackened backgrounds, multiplied. When anti-pollution laws allowed these forests' lichens to recover, peppery moths replaced their dark-bodied counterparts. Experiments showed that birds were less likely to find and eat moths that matched their backgrounds better. Thus moth-eating birds drove the spread and the retreat of dark peppered moths as pollution spread and retreated.

There are two lessons here. First, adaptation occurs when new alleles cause their bearers to multiply faster than the population's other members, thanks to new programs for some basic process. Second, other things being equal, natural selection favors alleles programming disguises that make their bearers harder for predators to find.

Many things had to happen, however, before complex organisms could respond adaptively to selection. How must genes be organized to allow adaptive evolution? How can such organization evolve?

Why Did Sexual Reproduction Evolve, and How Did This Happen?

Most of the animals we know come in two sexes. A female's eggs develop only if fertilized by a male's sperm. Similarly, a plant's ovules become viable seeds only if fertilized by pollen. For both, sexual reproduction is the ultimate goal in life.

Sexual reproduction presupposes meiosis, whereby diploid cells produce haploid gametes. Meiosis, which assigns each locus in a gamete one of its parent's two genes at that locus, involves an elaborate dance (fig. 8.2). Homologous chromosomes pair, locus by locus, and exchange homologous parts (an exchange called recombination) before a gamete receives one chromosome from each homologous pair. Meiosis enables orderly sexual reproduction. Bacteria and archaea are haploids that replicate their single chromosome when they divide; only eukaryotes have the elaborate dance of meiosis. How do eukaryotes benefit from orderly sexual reproduction with meiosis?

THE COSTS OF REPRODUCING SEXUALLY

Getting one's ovules fertilized, or fertilizing another's ovules or eggs with one's pollen or sperm, often involves a struggle. Many plants make flowers, some quite large, spectacularly beautiful or bizarrely shaped, or with striking scents, to entice animals to bring them pollen from other plants of their species. In many animals, males must either fight other males to obtain mates, or compete to attract them. Darwin saw this process as creating a distinct form of selection, sexual selection, where advantage is gained solely by winning from another male a mating with a female who would have mated anyway. He defined sexual selection as a selection that did not enhance a population's adaptedness to its way of life. Indeed, competing for mates is one cost of sexual reproduction.

Few contested Darwin's view that males often fight other males for mates. Indeed, in many species, males have elaborate structures such the horns of a ram, or a deer's antlers, to fight with rivals. The function of such structures is most obvious in deer, where males grow antlers for the mating season and shed them afterward.

Alfred Russel Wallace, the co-founder of evolutionary theory, however, doubted that males compete to *attract* females: could females really compare the attractiveness of different suitors? Only in 1980 did a graduate student, Michael Ryan, show that female preference affected male mating success in nature. He found that túngara frogs (their name imitates their call), *Physalaemus pustulosus,* were more attracted to male calls with more, deeper-toned chucks (the gara part of their túngara call). Indeed, indirect evidence of the importance of female preference is all around us for those with eyes to see or ears to hear. Cicadas fill tropical forests with the buzz-saw sound of their mating calls. Male peacocks and birds of paradise display elaborate ornaments that people of all cultures find beautiful, to attract mates (plate 8.1). In birds as different as manakins and bowerbirds, females exact elaborate and exhausting courtship rituals from their suitors. Female manakins demand elaborate dances with extremely rapid movements displaying vigor and exquisite balance and coordination. Female bowerbirds respond only to suitors that build large, strikingly decorated structures

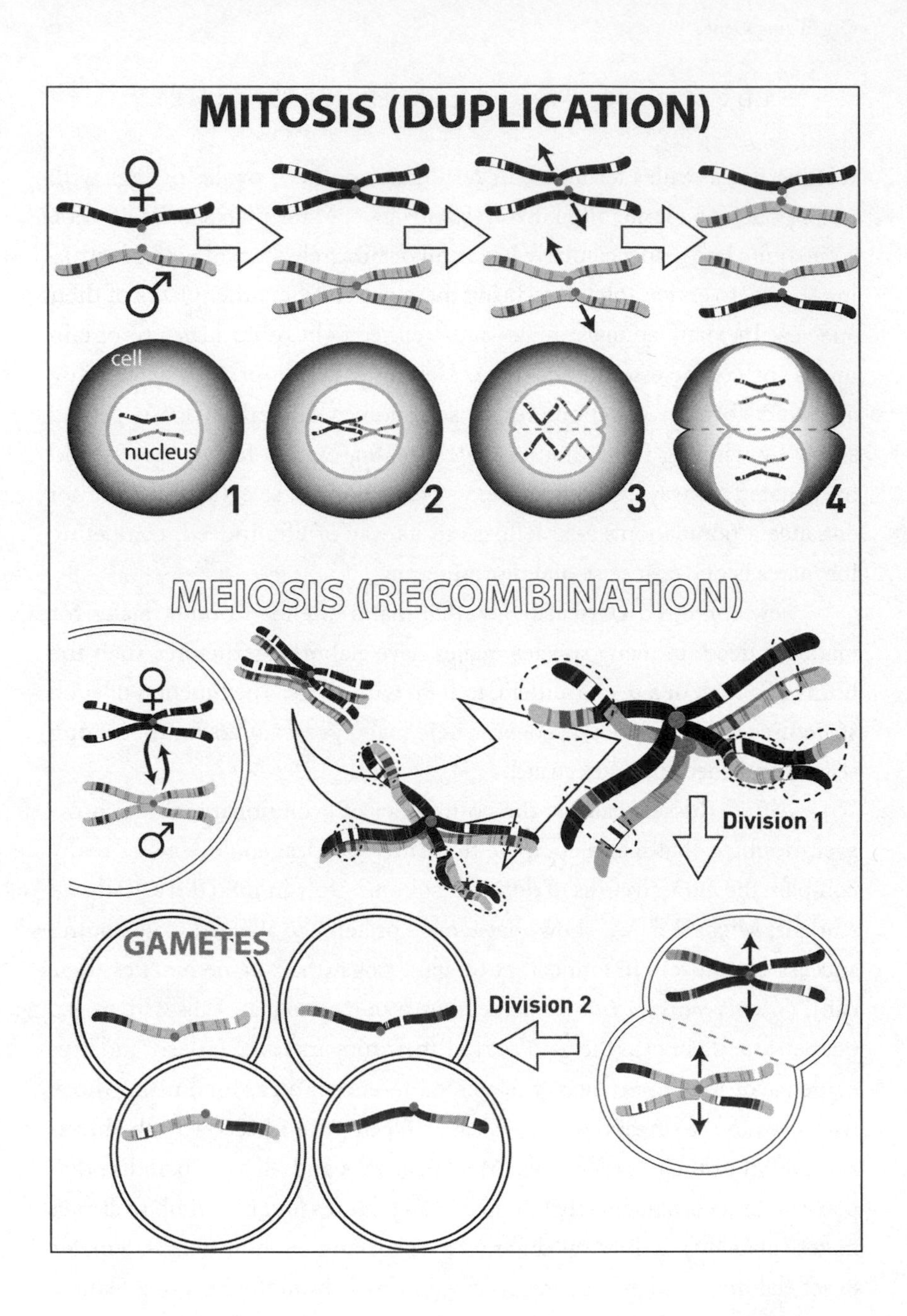

MITOSIS (DUPLICATION)
♀
♂
cell
nucleus
1
2
3
4
MEIOSIS (RECOMBINATION)
♀
♂
Division 1
GAMETES
Division 2

called bowers, sometimes with elaborate "avenues of approach." Suitors with enough vigor and stamina (and in manakins, capacity for rapid, well-coordinated acrobatics) to pass these tests might give their offspring good genes. But Darwin must have been right to conclude that females often choose the most beautiful and graceful males for the sake of their beauty, as if evolution were not always as bound by the useful or the profitable as a modern-day economist.

Even after insects join in copula, the male may stroke, "sing" to, caress, tap, or otherwise entice the female to eject the sperm of other males and let his sperm fertilize her eggs. When in copula, male insects use a whole Kama Sutra of techniques to persuade their mates to do their whole will, and females have many ways to avoid being imposed upon by methods that diminish the total number or quality of her offspring.

Although sexual reproduction is so eagerly sought, it reeks of paradox. First, every sexually produced organism begins life as a single cell, whereas asexual "vegetative reproduction," by runners or fragments of plants, corals, sponges, and the like, gives their young a safer start from a larger size. Second, sexual reproduction has a "50 percent cost." "Selfed"

Fig. 8.2. (*opposite*) Copying the genome for cell duplication (mitosis) in growth and maintenance is a very different business from assembling the haploid genome for gametes, eggs or sperm (meiosis). In mitosis, the maternal and paternal copy of each chromosome is duplicated, and each daughter cell gets one copy of each chromosome. Here, we consider only one pair of homologous chromosomes, and our diagram includes only a few of these chromosomes' 1,000+ genes. We figure one light chromosome from the father, and a homologous dark chromosome, with the same sequence of loci, from the mother. The genetically identical cells of a multicellular organism are produced from the fertilized egg by successive mitoses. Meiosis is the process whereby diploid organisms (whose cells have two genes at each locus) produce haploid gametes, sperm in males, eggs in females, with one gene at each locus. Here, each chromosome duplicates. The maternal chromosome pair adheres, locus for locus, to its complementary paternal pair and exchanges homologous segments in a process called recombination or crossing over. Recombination produces chromosomes with the same sets of loci but new combinations of alleles. Then two successive cell divisions produce four gametes, each with a single set of chromosomes. (Diagram by Damond Kyllo) plants, fertilized by their own pollen, produce young that inherit all their genes from their mother. Although a young produced by cross-pollination costs the mother as much as one produced by selfing, only half the genes of cross-pollinated plants are the mother's. Why, then, do so many plants avoid self-fertilization? Similarly, in most sexual animals, females bear, and sometimes rear, offspring, to which fathers give only genes, yet this mother's young inherit only half their genes from her. Why reproduce sexually? Does adding another's genes help mothers propagate their own?

THE OVERRIDING ADVANTAGE OF SEXUAL REPRODUCTION

In fact, sexual reproduction often favors adaptive evolution. An individual's program for its development and function is coded in its genes. The tropical botanist E. J. H. Corner likened a diploid individual's genes to a pair of football teams. Just as each team has eleven positions to fill—eleven functions to carry out—and one player, specially trained, for each position, so each haploid genome has thousands of loci, thousands of "positions," each governing a different process, with one gene at each locus. Each sexually reproduced diploid inherits one haploid genome from each parent. When it produces gametes, be they eggs or sperm, a cell's two genomes pair off, locus by locus. To form each gamete's haploid genome, a fair lottery chooses one gene at random from each locus. As the next chapter will explain, this ritual probably evolved so that undamaged DNA sequences could be paired with damaged counterparts to repair the latter. Nowadays, this ritual's function—indeed, the central purpose of sexual reproduction—is to test genes in different combinations, different genomes, some of which will be better than others. Corner remarked that fertilization and "meiosis, by shuffling the players, produces new teams, some of which will be better than others and more likely therefore to succeed in the game of life, while the poor teams are not only beaten but eliminated. Thus the sexual act of the union of gametes provides one of the means by which natural selection can improve the race of an organism."

In other words, without sexual reproduction, all of an individual's descendants inherit its genotype, mutations excepted. Here, a new allele's fate is governed less by its own effect on its bearers' reproduction than on how fast the genotype as a whole reproduces in which this allele first appeared, to which it is forever bound, just as in a rigidly stratified society a person's success depends far less on his ability than on his parents' status. Sexual reproduction with recombination shuffles a gene's descendants among many genotypes, so an allele's fate depends more on how it affects its bearers' fitness than on the quality of the genotype where it first occurred. Corner's shuffle produces both better teams replete with good players *and* worse teams with many bad players, enabling good players to join winning teams and bad players to be eliminated more quickly. Similarly, sexual reproduction produces more descendants with mostly good genes, which reproduce abundantly, and more descendants with many defective genes, which die. Sexual reproduction sorts good alleles from bad more effectively because it "levels the playing field," testing each allele more nearly on its own merits. Laboratory experiments confirm, moreover, that in yeast, sexual strains adapt faster than asexual ones to new conditions because in sexual strains, mutant alleles spread more nearly according to their own contributions to fitness.

Where the advantages outweigh the costs of sex, sexual reproduction serves the common interest of an organism's genes by testing each allele more nearly by its own effect on its genome's fitness. This is true even if an allele benefits some genotypes and harms others. In sexual organisms with abundant recombination, however, selection favors "good mixers" that enhance reproduction in all genotypes. Thus selection favors modular genetic systems. Some loci act nearly independently of others. Moreover, genetic systems are organized so that one segment, or one body part, can be modified nearly independently of others. This involves nearly unchangeable "character identification networks" that specify, say, an insect's forewings, and other genes that specify what form the character takes: in beetles, programming elytra that shield the resting beetle's hindwings; in diptera, programming wings a mosquito uses to fly. Modularity lets selection change one feature without affecting others adversely. This makes adaptive change much easier, especially in complex organisms. Thus orderly sexual reproduction favored genetic systems where selection distinguished good alleles from bad more easily, and effected adaptive change more quickly.

The advantage of sexual reproduction was tested by laboratory experiments with the much-studied millimeter-long nematode *Caenorhabditis elegans.* In one strain, nematodes reproduced only by fertilizing themselves; in another, nematodes reproduced only by cross-breeding. When exposed to a "mutagenic" chemical that increased mutation rate fourfold, the reproductive rate of the self-fertilizers dropped markedly, thanks to the accumulation of harmful mutations, whereas the outcrossers' reproductive rate did not change. When exposed to a disease-causing bacterium, the outcrossing strain evolved resistance to this disease within forty generations, whereas the self-fertilizers did not. Thus recombination let outcrossers adapt far better than self-fertilizers to adverse conditions. Indeed, sexual reproduction was a turning point in evolution: every many-celled animal, plant, and fungus is descended from sexually reproducing ancestors.

The primary bacterial analogue of sexual reproduction is a process called transformation (fig. 8.3). Strands of DNA drift in the water. Some kinds of bacteria absorb them. Some of their genes pair the absorbed DNA with homologous parts of their own chromosome, and exchange homologous genes. Here, a bit of the bacterial genome is "transformed" by DNA absorbed from outside. Transformation appears to be adaptive. Some bacteria absorb only strands of DNA from related strains, but, especially in life's earlier stages, bacteria and archaea exchanged genes with each other on a grand scale. Today, such "horizontal gene transfer" helps bacteria to evolve resistance quickly to powerful drugs, and is a major factor in their extraordinary chemical versatility. Homologues of genes enabling transformation play a crucial role in the meiosis of eukaryotes, which evolved just when their ancestors were domesticating the bacterial ancestors of mitochondria. Because meiosis pairs homologous chromosomes locus by locus, it favors an identical set of loci, identically arranged in chromosomes, within an interbreeding population, greatly reducing opportunities for horizontal gene transfer. Prokaryotes are noted for biochemical, eukaryotes for morphological, adaptation and diversity. Horizontal gene transfer enhances the first: does it hinder the second?

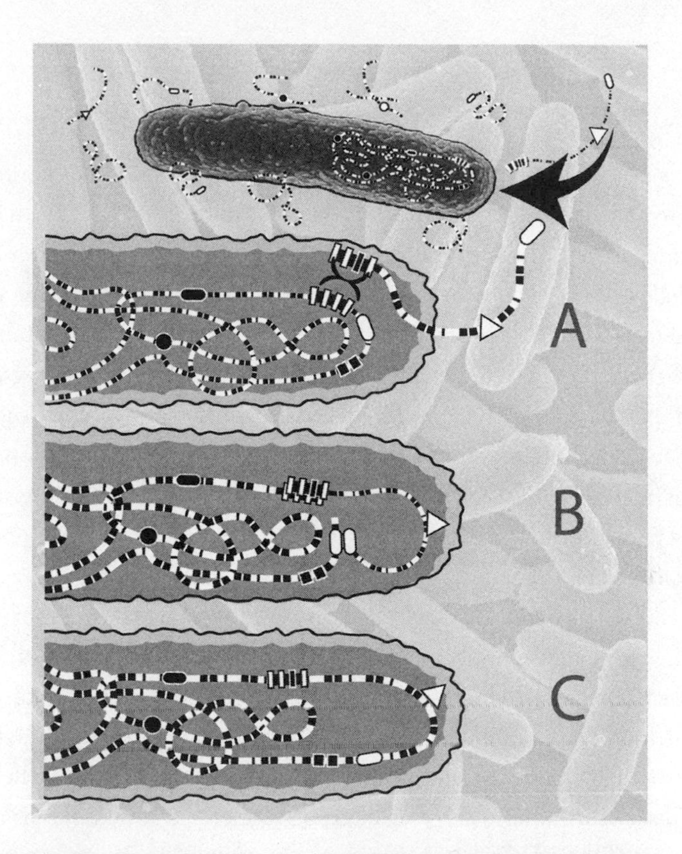

Fig. 8.3. Bacteria derive the advantage of recombination by absorbing stray genes from the surrounding water, which can exchange homologous sequences of DNA with the bacterium's own chromosome. This process can also introduce new genes, sometimes from very different species, into the bacterium, a process called horizontal gene transfer. Here, a segment of drifting DNA enters the cell, and one end of it pairs with a homologous gene on the bacterium's chromosome, and breaks this chromosome (A). The other end of the newly entered segment, which happens to be homologous to the free end of the bacterial chromosome, pairs with this end (B). The invasive segment is now integrated into the bacterial chromosome (C). (Diagram by Damond Kyllo)

What Makes Natural Selection on Selfish Genes Favor Adapted Individuals?

Meiosis offers opportunities for cheating. At most loci, the assignment of genes to gametes is a remarkably fair lottery where, at each locus, a gamete has equal chance of inheriting its grandmother's or its grandfather's gene. Mendel's laws hinge on the fairness of meiosis. The match of his experimental results with his predictions implies that meiosis is indeed fair. If, on the other hand, in heterozygous pea plants with both round and wrinkled alleles, pollen cells with a round allele "cheat" by somehow killing those with a wrinkled allele, round alleles would take over the population even if they harmed their bearers. Cheating disconnects how fast "cheater alleles" spread from how much they benefit their bearers, just as cheating on a test disconnects one's score from one's knowledge of the relevant subject.

Cheater alleles exist. They cheat in males, which make far more sperm than they normally need. For example, alleles controlling tail length in mice include **+**, the allele which, when homozygous, programs tails of normal length (long tails), an allele **T** that dominates **+**, where both **T+** heterozygotes and **TT** homozygotes program short tails, and various **t**-alleles that cheat by aborting sperm carrying alternative alleles, but do not bias meiosis in females (fig. 8.4). **Tt** heterozygotes have no tails, and 95 percent of the sperm of **Tt** males carry **t**. In contrast, **+t** heterozygotes have long tails, but 95 percent of the sperm produced by **+t** males also carry **t**. In males, this bias is measured by crossing **Tt** males with **++** females. If **t** and **T** had equal chances of entering each sperm, half the young would be short-tailed **T+**, and half, long-tailed **t+**. In fact, 95 percent of the young are long-tailed **t+**. Similarly, were meiosis fair, crossing **+t** males with **T+** females would yield one quarter long-tailed **++**, one quarter short-tailed **+T**, one quarter long-tailed **+t**, and one quarter tailless **tT** among the young. Of young with abnormal tails, 95 percent are tailless **+T** and 5 percent short-tailed **+T**, while among those with long tails, 95 percent are **+t** and 5 percent **++**. Thus, in both **tT** and **t+** males, most sperm carry **t**. Depending on which **t**-allele is involved, the cheater's spread is arrested either because all **tt** homozygotes die before birth or because male **tt** homozygotes are sterile. Thus no **t**-allele can completely take over a large population.

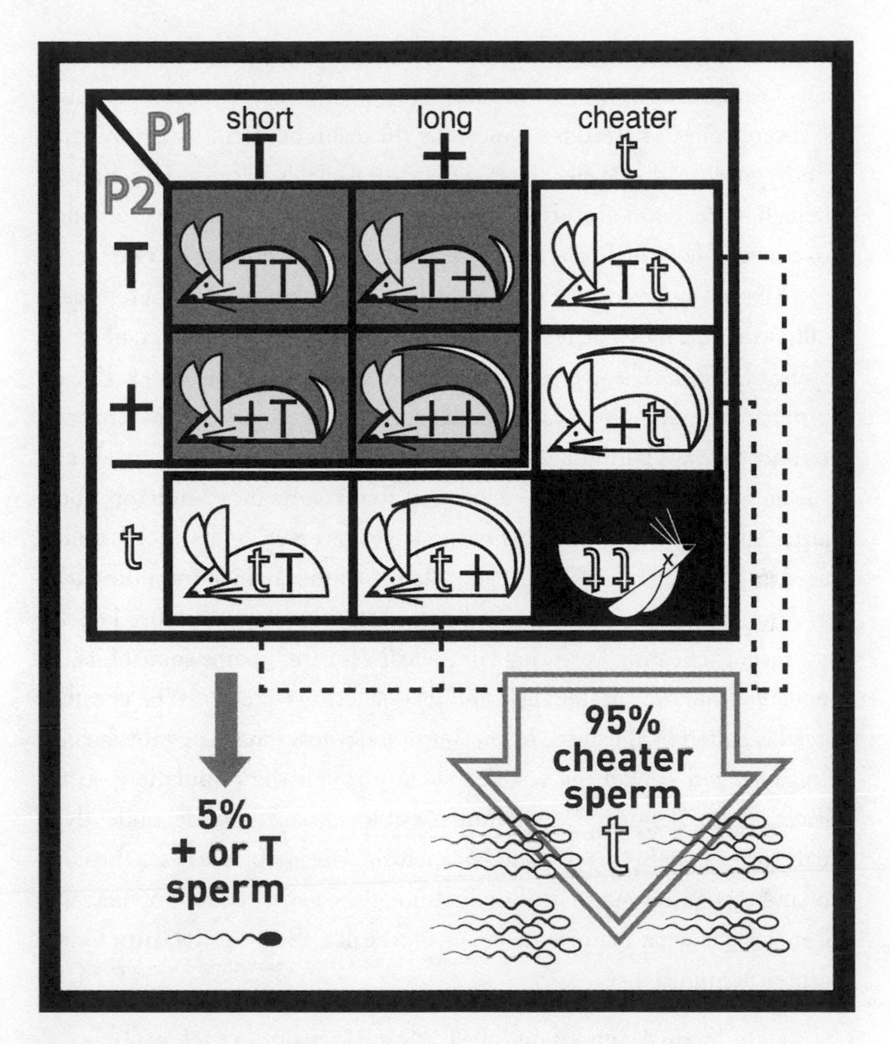

Fig. 8.4. Some genes spread by "biasing meiosis in their own favor" in male heterozygotes. In fact, these "cheater" alleles "bias meiosis" by disabling sperm carrying alternative alleles. This bias can spread harmful defects. The square shows the male genotypes formed in mice by the three alleles at the **T** locus: + (normal) programming long tails, **T** programming short tails, and cheater alleles, **t**. Males homozygous for **t** are either sterile or stillborn: they contribute no genes to future generations. In all males heterozygous for the **t** allele, 95 percent of the sperm carry **t**. If selection at unlinked loci did not keep biased meiosis so rare, evolution would often be maladaptive. (Diagram by Damond Kyllo)

Although alleles could easily spread by biased entry into sex cells, such cheating is very rare. To be sure, we normally observe such bias only if the cheater allele's spread is balanced by the death or sterility of individuals homozygous for this allele, as happens with **t**-alleles in mice. Other cheaters might spread too quickly through a population for biologists to notice. Nonetheless, few such balances are known. Why is cheating so rare?

Mendel's rules can be used to deduce that fair meiosis, whereby each of an individual's two genes for a basic process has equal chance of entering any of its sex cells, expresses the common interest of its genes. Phenotypic changes caused by the cheater, like other random phenotypic changes, are almost always harmful. Alleles at all loci suffer from a defect a cheater allele imposes on its bearers, but an allele that assorts independently of the cheater allele—that is to say, whose chances of entering a sex cell are not biased by the presence of the cheater allele—cannot benefit from the cheating. A mutant of such an independently assorting allele therefore benefits if it prevents cheating, restoring fair meiosis, for this spares some of its descendants from the cheater allele and its deleterious effects. If the cheater's spread is halted by the death of cheater homozygotes, restoring the fairness of meiosis will spread the restorer allele through the population, so the cheater allele disappears. As all mutants at loci assorting independently of the distorter benefit by restoring fair meiosis, fair meiosis reflects the common interest of the genes. This deduction of the genes' common interest in fair meiosis is what mathematicians would call a theorem. Its truth hinges on three assumptions:

1. Individuals inherit one allele at each locus from each parent at each locus.
2. A gene assorts independently of genes for most other basic processes.
3. Cheater genes are far more likely to spread harmful than helpful changes in basic processes.

Is this common interest effective? Is this common interest enforceable, and is it enforced? To find out requires more than logical deduction. Final answers await detailed studies of how cheater alleles work, and how they are suppressed, many of which are yet to be done. Here, I assume that this common interest has been effective, because

1. Although some alleles spread rapidly by securing biased entry into sex cells, most loci lack such cheaters. If biased entry into sex cells can happen, and can spread alleles so effectively, what keeps it from happening far more often?
2. The fairness of meiosis is a prerequisite for adaptive evolution, and adaptation is indeed the most striking characteristic of living beings.

Protecting the genes' common interest by ensuring fair meiosis is one of the most extraordinary achievements of evolution. The consequences of enforcing fair meiosis mimic Adam Smith's ideal economics. In both cases, competition obeying rules of fairness serves the common good, whereas unfair competition that violates these rules injures it. In both cases, the "collective," whether genome or society, must identify and suppress unfair competition. Judging by the prevalence of fair meiosis, genes have enforced rules of fair competition far more effectively than has any human society. Yet fair meiosis results from natural selection, an unconscious, mindless mechanism whereby miscopied alleles—miscopied instructions for some basic process—spread or disappear according to how well their changed instructions help their bearers survive and reproduce. That such a mechanism, incapable of planning or foresight, can bring forth and enforce a rule of fairness, a "moral code" that is the foundation of adaptive evolution, is truly a wonder of nature, which shows that interdependence, not intelligence, is the main driver of moral behavior.

In species with separate sexes, both diploid, whose members have many unrelated mates to choose from, devoting equal effort to raising male and female young serves the common interest of their genes. This is as true for sea lions, where the few largest, most pugnacious males monopolize nearly all the matings, as for monogamous birds. In all these populations, including ourselves, cows, and chickens, each individual has one mother and one father, so a successful sperm contributes as many genes to future generations as a successful egg. Because equal profits call forth equal effort to earn them, equal effort is devoted to raising each sex. Usually, as in human beings, it takes equal effort to raise a boy or a girl, and male births are as likely as female.

Fruit flies, *Drosophila melanogaster,* have four pairs of chromosomes. In three pairs, both members are alike. In the fourth pair, the members are unlike in males, which have a straight **X** and a bent **Y** chromosome, whereas females have two **X**'s. This pair of chromosomes is referred to as the sex chromosomes; the three other pairs are called autosomes. A sperm has equal chance of bearing an **X** or a **Y**, while all eggs carry **X**. Therefore, about half the newborns have one **X** and one **Y**, and are male; the others have two **X**'s, and are female. Only males carry **Y** chromosomes, so these chromosomes could, and sometimes do, spread themselves by programming male-biased sex ratios ("sex-chromosome distortion"). The difference between **X** and **Y** chromosomes, and the inheritance of **Y** chromosomes only through males, implies that only autosomal genes share a common interest in fair meiosis. As most genes are autosomal, theory predicts, and experiments with fruit flies confirm, that if sex ratio is male-biased, selection favors autosomal mutants programming more even sex ratios. Despite this "intragenomic conflict," sex ratio is nearly always even. More generally, meiosis is nearly always fair.

Even if sex ratio depends on temperature, the autosomal genome's common interest in an even sex ratio proves effective. The Atlantic silverside *Menidia menidia,* an annual fish, produces mostly female young if kept in cooler water, and mostly male young if kept in warmer water. How sex ratio depends on temperature varies with latitude, so that fish at all latitudes produce equal numbers of male and female young, in accord with their autosomal genes' common interest. D. Conover and D. Van Voorhees asked whether the common interest of these fish's genes could change their sex ratio. They placed some fish, and kept their descendants, in aquaria warm enough that the first fish produced mostly male young, and others in aquaria cool enough that the first fish produced mostly female young. After several generations, descendants were producing roughly as many male as female young in both the warm- and the cool-water aquaria.

This chapter began with alleles that spread by programming processes benefiting their bearers, leading to populations of individuals better adapted to their environments. For natural selection among genes to promote adaptation in complex organisms, however, genes must be organized in various ways. This organization is one of the outstanding achievements of evolution. This chapter discussed two central developments in eukaryotic organisms: the evolution of orderly sexual reproduction, and the evolution of fair meiosis. Sexual reproduction "leveled the playing field" for competition, allowing selection to test alleles more nearly by their own contribution to their bearers' fitness than by the accident of what genotype they were born in, and sift good from bad alleles more easily. Moreover, by promoting "good mixers," alleles which enhanced the fitness of all genotypes they occurred in, sexual reproduction promoted a "modular" organization where selection could change one characteristic without altering others. Anyone acquainted with governments incapable of legislating needed changes, because any change offends a plethora of special interests, will know how central modular organization is for promoting adaptive evolution in complex organisms. Finally, selection on genes, the ultimate units of self-interest, enforced a "moral rule" that served these genes' common good by allowing an allele to spread only by benefiting its bearers.

Genomes that facilitated adaptive evolution evolved when eukaryotes did. Yet rapid adaptive evolution and diversification awaited the evolution of animals, which was delayed for a billion years, not by oxygen shortage but by the difficulty of coordinating an accurate perception of the world with efficacious response to what is perceived.

This chapter showed how natural selection works, how it favors the genome's common good, and how it organizes genetic systems to favor adaptive evolution. Next we ask how natural selection favors diversification, social cooperation, and the pooling in mutualism of contrasting abilities of different species for their common good. Finally, we ask how the interplay among these processes brought forth evolution's most astounding and transformative achievements.

NINE

The Processes of Evolution

"[Evolutionary] improvement has been brought about in two main ways, which we may call *aggregation* and *individuation.* Individuation is the improvement of the separate unit, as seen, for example, in the series Hydra—Earth-worm—Frog—Man. Aggregation is the joining together of many separate units to form a super-unit, as when coral polyps unite to form a colony. This is often followed by division of labour among the various units, which of course is the beginning of individuation for the super-unit, the turning of a mere aggregation into an individual."

—J. B. S. Haldane and J. S. Huxley, 1927

INDIVIDUATION IS THE PROGRESSIVE adaptation of individuals. Aggregation involves either tightening cooperation and coordination within societies, as in the evolution of metazoans or honeybees, or among members of different species, as in the evolution of eukaryotes. How does selection favor diversification, social cooperation, and mutualism? How has interplay among these processes brought forth major evolutionary transitions?

Divergence and the Origin of Species

One species becomes two when a species is split into two parts and criteria for accepting mates diverges enough that no member of one part will mate with any of the other, or when, thanks to the substitution of incompatible alleles in these two populations, their hybrids cannot survive or reproduce. How can this happen?

Species sometimes arise "by accident." Accidental speciation is usually slow. Three million years ago, the Isthmus of Panama joined the

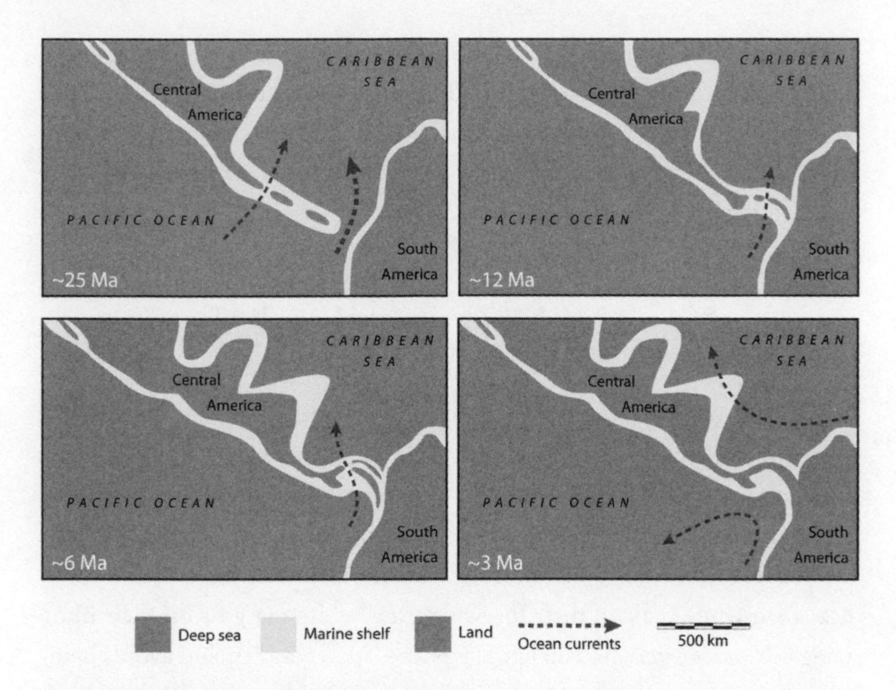

Fig. 9.1. Successive stages of growth of the isthmus of Panama, which split many marine populations that had lived in both oceans. Ma = millions of years ago. (Diagrams by Aaron O'Dea)

Americas, separating the Caribbean from the Pacific and splitting many marine populations (fig. 9.1). In most cases, the split halves became distinct species. In some species of snapping shrimp split by the isthmus, animals from different sides of Panama do not recognize each other as suitable mates, so that what was one species is now two. In a split population of long-spined sea urchins, *Diadema,* Pacific *Diadema* now spawn shortly after full moon, Caribbean *Diadema,* soon after new moon, so they cannot hybridize if reared together. On the other hand, gobies (nearshore fish), and some other sea urchins, separated by the isthmus for three million years, can still hybridize. When howler monkeys invaded Central America and Mexico after land connected the Americas, they split into two non-overlapping species that occupied a very similar range of habitat types, but did not interbreed. These howler species differed too little to invade each other's ranges: trade-offs played no role in their reproductive isolation.

More often, one species becomes two if its members face a trade-off whereby improving one ability diminishes another, especially if each ability is favored in a different habitat, whose residents specialize accordingly. Here, divergence in habitat or way of life and the consequent selection against hybrid matings cause speciation.

SPECIATION BY CHOOSING MATES WITH THE SAME WAY OF LIFE

Speciation often involves linking divergence into different ways of life with choosing mates that live in the same way. Speciation is easiest when this link is "ready-made." A million years ago, one species of poisonous passion-vine butterfly became two: the black, red, and yellow *Heliconius melpomene* (plate 9.1) split from the bluish black *Heliconius cydno* (plate 9.2) with white bars on their forewings and white margins on their hindwings. *H. melpomene* lives in light gaps and open country, and avoids being eaten by mimicking the poisonous open-country *H. erato* (plate 9.4); *melpomene* and *erato* coexist because their caterpillars live on different species of passionvine, *Passiflora. H. cydno* lives in the forest interior and mimics *H. sapho* (plate 9.3), which diverged from *H. erato* more than two million years ago, and, unlike *H. cydno,* has caterpillars specialized to a single species of passionvine. Speciation occurred because *Heliconius* almost always mate with others with similar wing colors, while predators eat hybrid *Heliconius* with unfamiliar wing colors. The link between divergence in color, which suits them for different habitats, and choosing mates with similar colors made speciation "magically" quick.

Plant speciation often involves invading a new habitat while avoiding mating with individuals adapted to the ancestral one. A flowering herb, the monkeyflower *Mimulus cardinalis,* with red flowers hummingbirds pollinate, growing below 2,000 meters altitude in California's Sierra Nevada, diverged from *Mimulus lewisii,* with pink flowers bumblebees pollinate, which grows above 1,600 meters. These plants face a trade-off between thriving at high versus low altitudes. Experiments show that in the lowlands, *Mimulus lewisii* fails where *M. cardinalis* succeeds because, unlike *cardinalis, lewisii* is vegetatively dormant during the winter. At high altitudes, *cardinalis* fails where *lewisii* succeeds because *cardinalis* is more susceptible to frost, and flowers too late to bear fruit before winter sets in.

Fertilizing flowers of one species with the other's pollen produces healthy, fertile hybrids: they diverged only recently. They do not interbreed in nature because they usually live in different places and because attracting bumblebee pollinators trades off against attracting hummingbirds. Douglas Schemske and H. Bradshaw, Jr., then professors at the University of Washington, demonstrated this by growing hybrids of these two species and crossing unrelated hybrids to produce a second hybrid generation, whose flowers showed all possible combinations of the floral characteristics of their parents' species—color, shape, size, nectar production, and so on. These hybrids were placed in a mountain meadow with both hummingbirds and bumblebees, to see which flowers attracted which pollinators. Redder flowers attracted more hummingbirds and fewer bumblebees, flowers producing more nectar attracted more hummingbirds, and larger flowers attracted more bumblebees. Moreover, the flower shape that lets hummingbirds fertilize their flowers and remove their pollen does not work for bees, and vice versa. Thus, shifting to hummingbird pollination allowed *Mimulus cardinalis* to evolve independently of the montane *M. lewisii* and adapt to warmer lowland settings. Here, speciation depended on the two *Mimulus* attracting different pollinators, which requires that each pollinator is able to recognize the flowers it can best use.

INNOVATION, HYBRID INVIABILITY, AND SPECIATION

Divergence between populations can involve genuine innovation. One such innovation enabled some montane sulphur butterflies, *Colias,* to invade warmer lowlands. Sulphur butterflies, so named for their yellow or orange wings, are a familiar summer sight in North America. Ward Watt, then a professor at Stanford University, who became interested in these butterflies in grade school, showed that to find food, mates, or suitable plants to lay eggs on, they must fly long and vigorously. As mentioned earlier, one genetic locus in these sulphurs programs an enzyme, phosphoglucose isomerase (PGI), that governs the supply of fuel to the flight muscles. Its function influences where, how long, and how vigorously its bearers can fly. In most populations, the PGI gene has two common alleles. Homozygotes face a trade-off: enzymes of the genotype that supplies fuel more rapidly are more likely to fail during hot weather. This enzyme has two similar, conjoined parts. In heterozygotes, PGI enzymes with one part from each allele have the best qualities of each allele, so heterozygotes can fly long, over a wide range of temperatures, with low risk of enzyme failure. The more heat-tolerant homozygotes, that supply fuel more slowly, are less competitive under cooler conditions. Heterozygotes survive longest and reproduce fastest. In lowland *Colias,* heterozygotes produce twice as many offspring as the homozygotes that supply fuel faster, and ten times as many as the homozygotes more resistant to warm weather. This heterozygote advantage maintains both alleles (and some rarer ones) in the population.

In North America, *Colias* sulphur butterflies all used to live high in the Rocky Mountains, or in Alaska or northern Canada. These cool habitats favor the ability to start flying early and move quickly over the ability to survive hot weather. These butterflies warm up in the morning by sitting with closed wings, broadside to the sun. Their hindwings are dark, as if brushed by coal dust, especially on their undersides and those inner edges closest to the body when the wings are closed, so they warm up faster.

Less than half a million years ago, *Colias* butterflies invaded the Great Plains and other hot lowlands, thanks largely to a mutation in the PGI gene that switched a single amino acid in the enzyme's sequence, which apparently made the mutant PGI enzymes less likely to fail in hot weather (fig. 9.2). Lowland sulphurs are also less darkly colored, reducing the risk of overheating. Lowland rarely mate with montane sulphurs, even in the laboratory, and their eggs are inviable. Because lowland and montane sulphurs live in different places and rarely meet, rarely mate if they do meet, and produce inviable eggs if they do mate, montane and lowland sulphurs have become different species.

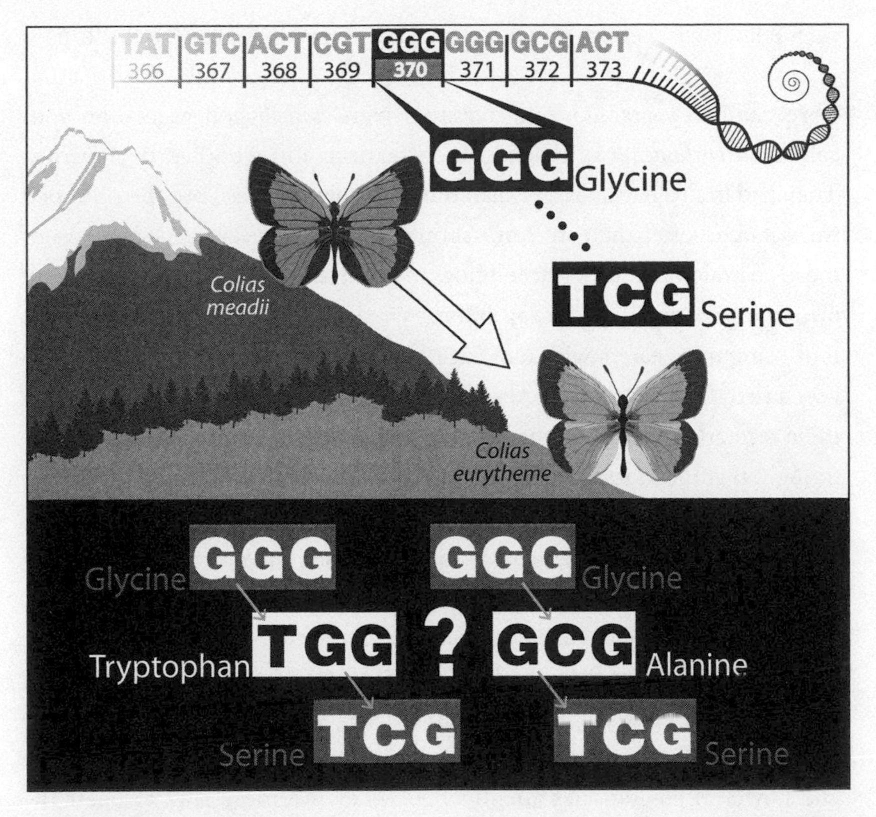

Fig. 9.2. How a single mutation can open a new way of life and facilitate speciation. *Colias meadii,* a butterfly of high altitudes in Colorado, with dark coloration that allows it to warm up more easily, and a brighter-colored lowland member of the same genus that absorbs less heat, *Colias eurytheme.* Phosphoglucose isomerase (PGI) controls the first chemical reaction involved in breaking down sugar to supply fuel for butterfly flight. These enzymes are made up of two nearly identical proteins, which face a trade-off: In homozygotes, the fastest catalysts are most sensitive to high temperature. Heterozygotes have the most functional enzymes, with one protein from a fast-catalyst allele, and one from a heat-tolerant allele. A mutation in one amino acid, involving two nucleotide changes appears to have programmed PGI enzymes far more tolerant of high temperature, which presumably played a crucial role in allowing their bearers to move from the cool highlands into the hot lowlands. (Diagram by Damond Kyllo)

SPECIATION DRIVEN BY BEHAVIORAL CHANGE

Three million years ago, our ancestors were two-legged vegetarian apes called *Australopithecus,* roaming the savannas and woodlands of Africa. They had brains hardly larger than those of chimpanzees, but they climbed trees much better than we can. About 2.6 million years ago, a group of these australopithecines learned how to make and use tools to kill and cut up large prey animals. By 2.3 million years ago, it was a new species of tool-using meat eaters with long arms and ape-like faces, distinct from all vegetarian *Australopithecus.* Although these meat eaters were no larger than their vegetarian ancestors, their brains were twice as large. Some biologists thought that their larger brains made them human, and named them *Homo habilis* for their use of tools.

Another, more obviously human species, *Homo erectus,* diverged from *Homo habilis* about 1.8 million years ago. They could walk and run, as we do today. Their brains were smaller than ours, but larger than those of *Homo habilis,* and their foreheads were lower than ours. Their mouths, jaws, and teeth, like ours, were much smaller than those of apes or *Homo habilis.* Chimpanzees have a bony crest atop the skull, to which their large and powerful jaw muscles are attached. They use these jaws to chew the tough, raw food they eat. *Homo erectus* lacked this crest, because their small, weak jaw muscles were attached to the sides, not the tops, of their skulls. Richard Wrangham, a Harvard anthropologist, deduced from the small jaws and teeth and weak jaw muscles of *Homo erectus* that they could not live only on raw food, which is much tougher and less digestible than if cooked (fig. 9.3). Food takes five times longer to chew raw than cooked, for tough raw food must be chewed into much finer fragments to be digested. Even so, our digestion can only extract 50–65 percent as much energy from raw food as from its cooked equivalent, and digesting raw food, even chewed so fine, takes more energy. Today's raw food eaters have a far wider range of choices in nearby markets than could ever be found in a day's walk through savanna. Nonetheless, living on raw food (as opposed to a vegetarian diet) makes it difficult or impossible for most women to conceive children. Wrangham inferred that *Homo erectus* began to evolve when a group

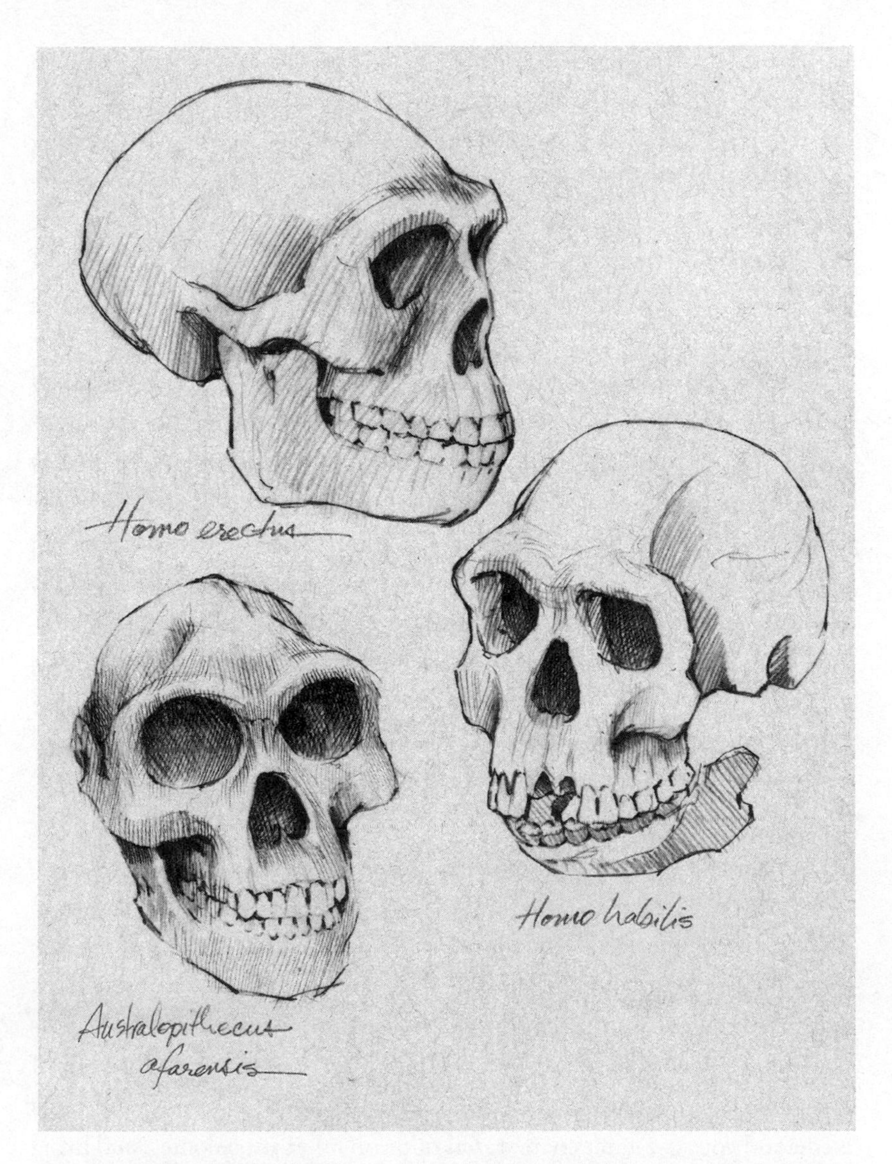

Fig. 9.3. Skulls of *Australopithecus, Homo habilis,* and *Homo erectus.* Using fire to cook food is thought to have enabled *Homo erectus* to chew food more easily. Even with smaller teeth and weaker jaw muscles, they could digest their food far more completely. (Drawing by Debby Cotter Kaspari)

of *Homo habilis* discovered fire and began cooking their food. Fire also scared off predators so effectively that the ability to climb trees, their former way of escaping predators, soon declined. Like aerobic respiration, cooking with fire extracted more energy, more quickly, from food, giving *Homo erectus* leisure to use their brains in more whimsical and imaginative ways. The coevolution of human brain and hand, however, opened so wide a range of abilities to *Homo sapiens* that other species of human being could not coexist with them.

Many think that evolutionary change is often, perhaps always, led by a new behavior. Terrestrial vertebrates began to evolve when their fishy ancestors regularly crawled onto land to find food. The evolution of whales must have begun when certain mammals started diving in shallow water for fish. How novel behavior triggers evolution is most evident in human beings. Learning to make and use tools allowed meat-eating *Homo habilis* to diverge from, and coexist with, their vegetarian australopithecine ancestors. Using fire to cook food allowed *Homo erectus* to diverge from *Homo habilis* and start on the path that eventually opened the whole world to their descendants.

Natural Selection, an Individual's Advantage, and Its Group's Good

Many animals benefit by living in groups. Insect-eating birds in a rain forest's understory join in "mixed flocks" to seek their food because a flock has more eyes to spot the approaching predator. Monkeys live in groups the better to detect and deal with predators and keep other groups out of the patch of forest that feeds them. In army ants group life allows the ants to defend young, mount coordinated raids on other ant colonies, and bring food they catch back to the nest (where living ant workers form the walls). No lone ant could do such things. The problem of group life, however, is how to ensure that most or all of a group's members cooperate by contributing to the good that benefits all.

Plate 5.24. *Selaginella* on the forest floor, Lambir Hills, Sarawak, Borneo. It produces both megaspores and microspores, a first step toward evolving seeds. Its ancestors diverged from the lineage leading to ferns, conifers, and flowering plants about 410 million years ago. This lineage once included tall trees, but only herbs, such as club mosses and *Selaginella,* survive. (Photograph by Christian Ziegler)

Plate 5.25. Flowering plants can maintain genetic variation even when made rare by their pests thanks to pollinators that travel long distances. Here, a hummingbird, the long-tailed hermit *Phaethornis superciliosus,* is pollinating a passionflower, *Passiflora vitifolia.* (Photograph by Christian Ziegler)

Plate 5.26. A cycad leaf, with many leaflets. A leaflet's only vein is its midrib. Cycads cannot transport water very effectively: they are now slow-growing members of the forest understory. (Photograph by Christian Ziegler)

Plate 5.27. Part of the leaf of the fast-growing fig tree *Ficus insipida,* whose dense venation allows effective transport of water to all parts of the leaf, enhancing its rate of transpiration and photosynthesis. A hundred million years ago, evolving high vein density allowed flowering plants to diversify and begin to dominate most forests. (Photograph by Christian Ziegler)

Plate 6.1. Unaccustomed disturbance usually reduces diversity. Here, human activity has replaced diverse forest with monotonous grassland in Panama. Intact forest remains behind the paja grassland. (Photograph by Christian Ziegler)

Plate 6.2. Leaves need light for photosynthesis, and in mature forest light is divided most unevenly among leaves. Here, deeply shaded forest floor on Barro Colorado Island, Panama, supports few leaves. (Photograph by Christian Ziegler)

Plate 6.3. On this same island, the forest canopy is crowded with leaves. Here, the yellow crowns belong to *Tabebuia guayacan,* the blue ones to *Jacaranda copaia.* (Photograph by Christian Ziegler)

Plate 6.4. Northern forests are inhospitable to animals during the winter. They are cold, and offer little fit to eat. A leafless forest on Ruegen Island, off the Baltic coast of Germany, in bleak midwinter. (Photograph by Christian Ziegler)

Plate 6.5. Plants need light: some plants "cheat" to get it. Here, epiphytes in Thailand, staghorn ferns (*Platycerium* sp.), are taking advantage of a tree's sunny branch to get their sunlight. (Photograph by Christian Ziegler)

Plate 6.6. Giant flower of a plant, *Rafflesia,* that parasitizes the roots of vines, *Tetrastigma,* in the grape family, Vitaceae. This flower attracts pollinating carrion-flies by seeming to offer fresh carrion to eat and to lay eggs in. (Photograph by Christian Ziegler)

Plate 6.7. Dependence on different habitats. This giant damselfly, *Megaloprepus caerulescens,* whose young live in water and adults on land, is defending a water-filled tree-hole (not shown) that attracts females to mate with him so they can lay eggs in this hole. Nymphs hatch from these eggs and grow and mature in there. (Photograph by Christian Ziegler)

Plate 6.8. Migrating sulphur butterflies, *Phoebis* sp., presumably seeking young leaves on which to lay eggs, feeding on fig fruit on the banks of the Manu River in southeastern Amazonian Peru. (Photograph by Christian Ziegler)

Plate 6.9. A protector of the forest at work: *Micronycteris microtis,* a small Panamanian bat that eats large insects, eating a katydid. (Photograph by Christian Ziegler)

Plate 6.10. Orchid bees, *Euglossa* sp., attracted by the sweet smell of the flower of an orchid, *Gongora powelli,* in Panama. Male orchid bees are effective pollinators: they fly long distances to gather various odors which they mix into a perfume to attract females of their species. (Photograph by Christian Ziegler)

Plate 6.11. A newly eclosed fig-pollinating wasp (*Pegoscapus hoffmeyeri)* emerging from a fig "fruit," a flowerhead turned outside in, of the fig tree *Ficus obtusifolia.* These wasps have been domesticated to serve as this fig species's pollinators. They live only three days and mostly drift with the wind, but they often pollinate trees more than 10 kilometers from their natal tree whose pollen they carry. (Photograph by Christian Ziegler)

Plate 6.12. Patch reef, from Raja Ampat, which Alfred Russel Wallace visited to see birds of paradise, in West Papua (the west end of New Guinea). Coral and mollusk diversity are higher in this region than in any other marine setting. Left, *Acropora* table corals, with feeding soft corals to their right; above left, back of the table corals, *Acropora* branching corals. An unidentified branching coral is growing from the table coral. An unidentified red sponge is near the center. (Photograph © Robert Delfs, 2017)

Plate 6.13. Badlands, formed from strata of successive fossil soils, Petrified Forest National Park, Arizona. (Photograph by Wil Meinderts/Buiden-beeld/Minden pictures)

Plate 6.14. Interdependence: A seasonal local migrant, *Artibeus lituratus*. Many members of this species fly to a different habitat during seasons when Barro Colorado Island is short of suitable fruit. (Photograph by Christian Ziegler)

Plate 6.15. Agouti, *Dasyprocta punctata,* with orange fruit of the spiny palm *Astrocaryum standleyanum.* Agoutis bury the seeds of many trees, thus protecting them from insect attack: they eat buried seeds when fruit is scarce. On islands too small to support agoutis, this service is not available, so some tree species die out. (Photograph by Christian Ziegler)

Plate 6.16. The impact on small islands of competitive invaders: a dense thicket of the exotic invader *Psidium cattleyanum,* strawberry guava, in a Malagasy treefall gap. Such dense stands of strawberry guava are crowding out the less shade-tolerant native vegetation in Mauritius, a much smaller island. (Photograph by Christian Ziegler)

Plate 8.1. Sexual selection: This male peacock in Thailand has long brightly colored feathers which he displays to entice females, who appreciate such beauty, to mate with him. Human beings also find these displays beautiful. As Darwin said, a female bird's aesthetic is rather like our own. (Photograph by Christian Ziegler)

Plate 9.1. *Heliconius melpomene,* which mimics *H. erato,* near its caterpillar's food plant, *Passiflora menispermifolia,* whose leaves already bear a few small yellow eggs that it or a competitor has laid. *H. melpomene* and *H. cydno* diverged less than a million years ago. *H. melpomene,* like *H. erato,* are open-country butterflies. If different bad-tasting butterflies share the same warning colors, they can split the cost of teaching predators to avoid them. (Photograph by Christian Ziegler)

Plate 9.2. Different habitats favor different colors. *Heliconius cydno,* shown here, mimics *H. sapho:* they are shade-loving butterflies. They speciated readily because butterflies like to mate with those of similar colors; *erato* and *melpomene* diverged long enough ago that they find each other unattractive for other reasons, as is the case with *cydno* and *sapho.* (Photograph by Christian Ziegler)

Plate 9.3. *Heliconius sapho,* which *H. cydno* mimics. (Photograph by Christian Ziegler)

Plate 9.4. *Heliconius erato,* alighting on its caterpillar's food plant, *Passiflora biflora.* (Photograph by Christian Ziegler)

Plate 9.5. A white-faced monkey, *Cebus capucinus,* eating a flower of balsa (*Ochroma pyramidale*) in Panama. These monkeys cooperate to fight off other groups. A group's monkeys share a common interest in their group's size and cooperativeness: the more members join a fight with another group, the more likely the first group is to win. More members join a fight if it is closer to the center of their home range. (Photograph by Christian Ziegler)

Plate 9.6. Brief-exchange mutualism: A butterfly, the distasteful *Heliconius hecale melicerta,* pollinating a flower, *Cephaelis,* called hotlips. In such "brief-exchange mutualisms" many animals cheat by robbing nectar or pollen without taking the pollen to fertilize other plants. Plants pay this price for a badly needed service. (Photograph by Christian Ziegler)

Plate 9.7. Brief-exchange mutualism and cheating plants. Orchids are past masters at tricking insects into pollinating them without giving them any reward. A deceptive, nectarless pansey orchid, *Diuris magnifica,* in Western Australia cheats insects by mimicking a nectar-rich pea flower that rewards pollinators. (Photograph by Christian Ziegler)

Plate 9.8. This white-necked Jacobin hummingbird, *Florisuga mellivora,* is stealing nectar from a balsa flower, *Ochroma pyramidale,* without pollinating the flower or taking its pollen to fertilize a flower on another balsa tree. (Photograph by Christian Ziegler)

Plate 9.9. Polyps of the reef coral *Montastrea cavernosa,* which is provided carbohydrates by symbiotic algae, zooxanthellae, in return for mineral nutrients and a safe, sunlit home. The coral maintains mutualism by expelling non-performing symbionts. (Photograph by J. L. Wulff)

Plate 9.10. Mutualistic reef sponges, *Iotrochota birotulata* (green), *Amphimedon compressa* (crimson), and *Aplysina fulva* (ochre) off Panama's Caribbean coast, August 2011. They grow intertwined, allowing one sponge to benefit from another sponge species's ability to resist different diseases, predators, or other hazards. (Photograph by J. L. Wulff)

Plate 10.1. Ring-tailed lemurs, *Lemur catta,* on a tree branch in southwestern Madagascar. Although these lemurs are far poorer than monkeys at solving human-devised "learning puzzles," they live in stable social groups. They must learn the social skills needed to do so. Working on these lemurs gave Alison Jolly the idea that social life is the primary force driving the evolution of primate intelligence. (Photograph by Teague O'Mara)

A CHALLENGE OF GROUP LIFE: HOW TO COPE WITH CHEATERS?

A social animal's fellow group members are its closest competitors for food, mates, and shelter. What keeps competition among a group's members from degenerating into a cheating contest that destroys all the good of cooperating? What factors create, and what circumstances allow enforcement of, a common interest among them in cooperating? How to prevent cheaters from annihilating the benefits of social life is the central problem of social behavior, one that all group-living animals have solved. This problem has analogues at all levels of biology and even beyond.

For example, each of us, like other animals from flies to elephants, is made of cells that can multiply on their own. We each start as a fertilized egg that multiplies by dividing. This multiplication gives rise in superbly ordered fashion to all the kinds of cells, tissues, and organs that make up a human body, each in its proper place, with its proper function. Normally, the coordinated activities of these cells, by which we live, move, eat, think, and act, far surpasses the intricate coordination of the different activities of bees in a honeybee hive. Yet a rogue cell's multiplication sometimes escapes control, causing a cancer that may kill the person who has it. How can human beings, or other animals, escape cancer?

Similarly, human beings are social animals: we depend on each other in countless ways. As Aristotle realized, we cannot live except within some society. Yet a human economy is an arena of competition as well as a web of interdependence: we compete for resources in limited supply. Philosophers, economists, and statesmen from Plato and Aristotle onward have struggled to learn how best to reconcile the self-interest of individual human beings with the good of their society.

TRAGEDIES OF THE COMMONS

In human groups, "tragedies of the commons" can arise through conduct that gives a person an advantage over his neighbors but hurts the group if all do likewise. For example, in a village of sheepmen holding their pastureland in common, a villager profits by placing extra sheep on the commons. If all do so, however, overgrazing ruins the commons and all suffer. Again, someone in an otherwise honest society may benefit, at least briefly, by cheating on his contracts, but a society where contracts are meaningless cannot function. Finally, a worldwide tragedy is developing because using more energy makes a person's life much easier and more agreeable, and grants access to many more occupations. Using this energy, however, releases carbon dioxide, which overheats the world, melts ice caps, and floods lowlands, injuring us all.

In self-governing traditional villages, residents avert such tragedies by enforcing mutually agreed rules on how many animals each can pasture on the commons, which presupposes the ability to devise and communicate, and the willingness to implement, what needs to be agreed upon. Outsiders, who do not share the villagers' common interest in sustainable exploitation, are excluded from the commons.

In large, well-governed societies, contracts are legally enforceable and cheaters are punished. Nonetheless, enforcing the common good by either communal action of local societies, a nation's legal power, or international agreement is risky and uncertain. Governments may deprive villagers of the right to manage their commons or to exclude outsiders, allowing overexploitation or takeover by nonresidents. A nation's legal power may serve influential individuals rather than the common good. Legislators may not know what serves the common good, or how to enforce it. Nations may disagree on how to share the burdens of achieving the common good. Talks on reducing carbon dioxide emissions, for example, have faltered, because the immediate inconvenience of these reductions outweighs concern for the future. Can other animals cope better with the problems and risks of group life?

COOPERATING WITH RELATIVES: KIN SELECTION

Some social animals cooperate only with relatives. In wasps and bees that make long-lasting nests, an individual sometimes stays at her mother's nest and helps feed and care for her mother's or sister's young. Helping a successful relative reproduce serves her genes' common interest if more copies of her genes (more precisely, more genes inherited unchanged by her *and* her relative from a recent common ancestor) are passed on to the next generation by helping that relative reproduce than by bearing young of her own. The nocturnal bee *Megalopta genalis,* which gathers pollen for an hour after sunset and an hour before dawn, when there is enough light for its specially adapted night-vision eyes to see, shows social life at its simplest. Although these bees often nest alone, two bees form a society, the smallest one possible, when a bee starves her daughter by feeding her less. Consequently, to pass on any genes at all, the starved daughter must stay and help her mother. She gathers pollen, enabling her mother to lay more eggs and defend the nest against marauding ants. For these bees, living alone, and forcing daughters to become non-reproductive workers, yields equal numbers of young.

In the tropical wasp *Polistes canadensis,* social life is essential for success (fig. 9.4). Their nests—"combs" of hexagonal cells, many filled with brood or young, with adult defenders on the surface—often hang from the eaves of neglected buildings. In natural forests their nests hang from tree branches. A successful colony begins with a single fertilized female. It grows into a nest with hundreds of wasps—a reproductive queen (the largest nests have two or more), foragers, defenders, caretakers, and more. How a wasp colony with division of labor develops from a single fertilized queen resembles how an animal with many types of cells develops from a single fertilized egg. Predators, however, destroy an entire brood often enough that workers move among nests of related queens, so that some of their work will propagate their genes. True, naked force helps to hold a wasp colony together. A nest defender may kill and replace the queen, just as in the Roman Empire generals sometimes rebelled and replaced their emperors. Defenders, in turn, may force other wasps to forage, and so forth. Yet the cooperation and division of labor among a colony's wasps is ultimately driven by the fact that when workers help their mother or sister reproduce, they pass on some of their genes to the next generation.

Honeybee colonies derive many advantages from living in groups:

Fig. 9.4. A new stage of social complexity: a *Polistes canadensis* nest with wasps on it. This colony is too large for the queen to forcibly dictate division of labor: the workers must benefit from helping relatives reproduce. (Photograph by Patrick Kennedy)

they can fend off predators and find food more effectively thanks to the intricate division of labor among workers for helping their queen reproduce. A worker benefits from group life because, even if she dies early, the help she gave the queen while alive spreads the genes she shares with her. How do honeybee colonies prevent cheating by "free-riders," who benefit from their group's efforts without contributing to them? Here, kin selection plays a key role. Some workers, about one in ten thousand, produce male offspring and do nothing to help the queen, their mother, reproduce. The queen controls such cheating by creating a common interest among her workers in helping her. The queen mates with 16 to 20 males and mixes the sperm thoroughly. A worker eats another worker's egg because, in natural colonies, it would usually be a half sister's, thus less closely related to her than a queen's egg. Even though the three or four egg-laying workers in a colony of 35,000 lay 7 percent of the colony's unfertilized eggs, other workers eat 99 percent of these eggs, so only one in every thousand of a colony's males has a worker mother. This mutual policing enforces a common interest among a queen's workers in helping her. This enforcement is so effective that honeybee societies have become extraordinarily complex and their cooperative behavior wonderfully adapted to the opportunities and problems they face.

HOW AND WHY DO UNRELATED ANIMALS SOMETIMES COOPERATE FOR THEIR COMMON GOOD?

Interdependence, not intelligence, is the true mother of "moral rules." Because each gene's reproduction depends on the functions of all the others, a genome's genes have a common interest in enforcing fair meiosis, which ensures that alleles spread only if they promote their genome's common good. Fair meiosis is a basic moral rule that was spread and is maintained by the mindless mechanism of natural selection.

Over two millennia ago, Plato (Republic, Book I, 352c) realized that a group of thieves, depending on one another to rob successfully, must "hang together or hang separately." These thieves can only work together effectively if they deal justly with each other. Two centuries later, a Chinese philosopher expressed similar views:

> Once a follower of the great brigand Chih asked him whether thieves have any use for wisdom and morality. "To be sure, they do," said the brigand Chih, "just as much as other people. To find oneself in a strange house and guess unerringly where its treasures lie, this surely needs Inspiration. To be the first to enter needs Courage, to be the last to leave needs Sense of Duty. Never to attempt the impossible needs Wisdom. To divide the spoil fairly needs Goodness. Never has there been or could be anyone who lacked these virtues and yet became a really great brigand."

Many birds and mammals depend on fellow group members to help ward off predators and competitors. Greater anis, *Crotophaga major,* are large black birds of the cuckoo family. They live near Barro Colorado's shore and nest in groups. They pair off, but two or more pairs lay eggs in one nest, because lone pairs cannot prevent predators from eating eggs or young. Safe nesting sites are scarce, and the more pairs, the better defended the nest, so more than two pairs may share a nest in a safe site. As a graduate student, Christina Riehl found that all of a nest's pairs raise similar numbers of chicks. Females equalize reproductive success by ejecting eggs of others from the communal nest until they themselves begin to lay. After laying an egg herself, a female stops removing eggs. Once the last female lays an egg, ejection of eggs ceases, and females lay in turn. After all have laid their eggs, incubation begins. This behavior enforces the equitable reproduction and synchronous hatching needed to create a common interest in nesting together. Newly laid eggs are covered with a chalky white coating, which slowly wears off, progressively revealing the blue eggshell beneath. This feature makes it easy to detect eggs laid by females from other nests after incubation begins.

When a group's members need each other to survive, they share a common interest in their group's effectiveness. Group-living monkeys need each other to help ward off predators and repel competitors. When following radio-tagged harpy eagles released on Barro Colorado Island, Janeene Touchton found that white-faced monkeys, *Cebus capucinus,* coordinated remarkable evasive/defensive tactics against these eagles (plate 9.5). When they saw one approaching, male *Cebus* ran to the tree crown, waving their fists threateningly at it, while their troop mates dropped to the base of the tree crown, quickly descended the tree, and ran off, followed by the defending males. Another researcher, Margaret Crofoot, has seen these monkeys mob ocelots, which sometimes eat monkeys they surprise on the ground. If they see the ocelot first, they mob it, forcing it to abandon its hunt, and sometimes putting it in real fear. A white-faced monkey group fights neighboring groups to defend and expand its territory's boundaries. Crofoot finds that a monkey group nearer the center of its territory is more likely to win a fight: victory matters more to all of its members, so more of them join the fight, creating a "home-court advantage." Because group members depend on one another to survive, a monkey endangers its own future if its negligence causes a fellow member's injury or death, or if it so limits another monkey's access to food that the victim no longer benefits from being in that group. A group's monkeys thus share a common interest in each other's welfare, and, like members of a robber gang, must treat each other accordingly.

The common interest of chimpanzees in their group's welfare leads them to enforce some rudiments of morality. Chimpanzees can recognize each other. They can picture how they would behave were they in another chimp's place, an ability that is most obvious in how they deceive each other. A young chimp sometimes leads adults toward one source of food so he can sneak off and feed at another, undisturbed by his more powerful elders. Their ability to imagine themselves in another chimp's shoes leads chimps to expect others to do unto them as they do unto these others, and they normally do unto others as they have been done by. Chimps that fight are expected to reconcile afterward. A chimp that conspicuously refuses a gesture of reconciliation risks attack by its whole group, which benefits from harmony among its members. If a dominant alpha male unduly favors allies or favorite consorts instead of settling disputes among other group members fairly, it may be dethroned through joint action by other group members in favor of one more inclined to do justice. These shared standards of behavior amount to a proto-morality that shines through the extraordinary mix of force and deception that sparks so many interactions among a group's chimps. This proto-morality strengthens the common interest of the chimps in their group's welfare by giving each useful member a greater stake in it.

Cooperation Among Members of Different Species: Mutualism

In complex human societies, economic productivity, diversity of occupations, and social welfare hinge on cooperative endeavor. Mutualism, where members of two or more species cooperate for their mutual benefit, is equally essential for the productivity and diversity of ecological communities. Even when they first evolved, land plants needed mycorrhizae, root fungi, to extract nutrients from the soil in return for starches from the plant. Before red algae had diverged from green, their ancestors had transformed live-in cyanobacteria into chloroplasts, enabling the evolution of many-celled plants.

As with members of the same species, cooperation among different species evolves only if members of each partner species benefit from helping the other—that is, if cooperation serves their common good. Because members of different species are never related, kin selection cannot favor mutualism between species. Although a non-reproductive ant worker spreads her genes by helping her mother reproduce, Darwin said that discovering an individual that ceased reproduction to benefit another species would destroy his theory. Most organisms, however, compete less intensely with members of other species than those of their own: Darwin knew that more distantly related species, whose needs differ more, coexist more readily. Moreover, distantly related species more often have complementary abilities that can be pooled for mutual benefit. The most famous mutualisms—plants and their pollinators and mycorrhizae, corals and their zooxanthellae—involve members of different kingdoms.

Mixed-species bird flocks forage together for safety from predators. Dietary and behavioral differences among these species reduce the costs of flocking together. As students, Russell Greenberg and Judy Gradwohl found that dot-winged antwrens, *Microrhapias quixensis,* and checker-throated antwrens, *Epinecrophylla fulviventris,* in Panama move through the forest understory together, making the rounds of their jointly held territory. Dot-winged antwrens glean insects from live leaves, and checker-throated antwrens glean insects from wads of dead leaves. They both focus on vine tangles. Slaty antshrikes, *Thamnophilus punctatus,* which have smaller territories and sally long distances after prey, join the flock when it moves through. White-flanked antwrens, *Myrmotherula axillaris,* have territories including two or three mixed flocks. They tend to join a flock when it forages away from vine tangles. Differences in hunting techniques or foraging sites allow these species to flock together with little competitive interference.

Defending Cooperators Against Cheaters

Cooperators always risk being cheated by partners. Cooperation normally serves the common good only if cheating can be restricted. How are mutualisms defended against cheating? To answer, we distinguish two types of mutualism. One type involves brief exchanges, as when an animal disperses a tree's seed or pollinates its flower just once, like a customer visiting a distant shop (plate 9.6). The other is symbioses, where one partner spends more than half its life linked to the other, like the fungus that leaf-cutter ants cultivate to digest leaf fragments.

BRIEF-EXCHANGE MUTUALISMS

Plants and the animals pollinating them often cheat each other. Many orchids attract pollinators without providing food in return for service (plate 9.7). Some even make a flower part that mimics the odor of a female of their insect pollinator, so it tries to copulate with this flower part, gathering pollen in the process. Birds or bees may pierce a tree's flowers at the base to get nectar, rather than extracting it in a way that dusts them with pollen (plate 9.8). Parrots eat rather than disperse the seeds in a tree's fruit. Evolutionary change is often the plants' only response: plants whose flowers and fruits attract only animals likely to pollinate them or disperse their seeds have more young. Despite these cheaters, employing animals as pollinators and seed dispersers makes flowering plants better competitors, and keeps tropical forest diverse and productive.

SYMBIOSES: COOPERATORS THAT LIVE TOGETHER

The story is quite different for symbioses. Here, cheating is severely limited, and we know how it is controlled. Cheating is rarest in symbionts passed from mothers to offspring, because these symbionts reproduce only if their hosts do. Leaf-cutter ants domesticated the miracle fungus that digests so many kinds of leaves for them by transmitting it only from mother to daughter. Each new queen takes a mouthful of her mother's fungus garden to start her own. Later, her workers will exclude contaminants and keep the fungal strain pure. Thus the fungus reproduces only if the ants do. Domesticating this fungus enabled its host ants to become voracious leaf eaters. Similarly, mitochondria (and chloroplasts) were tamed by passing them only from mother to daughter cell, and from only one parent to sexually produced offspring.

Nematodes, roundworms that parasitize fig-pollinating wasps, reveal how inheriting symbionts only from one's mother enforces mutualism. To rehearse these wasps' life cycle, a certain number of fertilized female wasps, characteristic of the fig species, enter a fig syconium, pollinate the flowers inside this perforated ball, and lay eggs in half of them. When these young mature, they mate, and fertilized females fly off to find other fig trees ready to pollinate (fig. 9.5). Allen Herre, who began studying fig wasps on Barro Colorado as an undergraduate, found that fig-pollinating wasps carry parasitic nematodes. These nematodes eat their hosts after these hosts reproduce, and then mate and lay eggs. Young nematodes enter young female wasps just before they fly away.

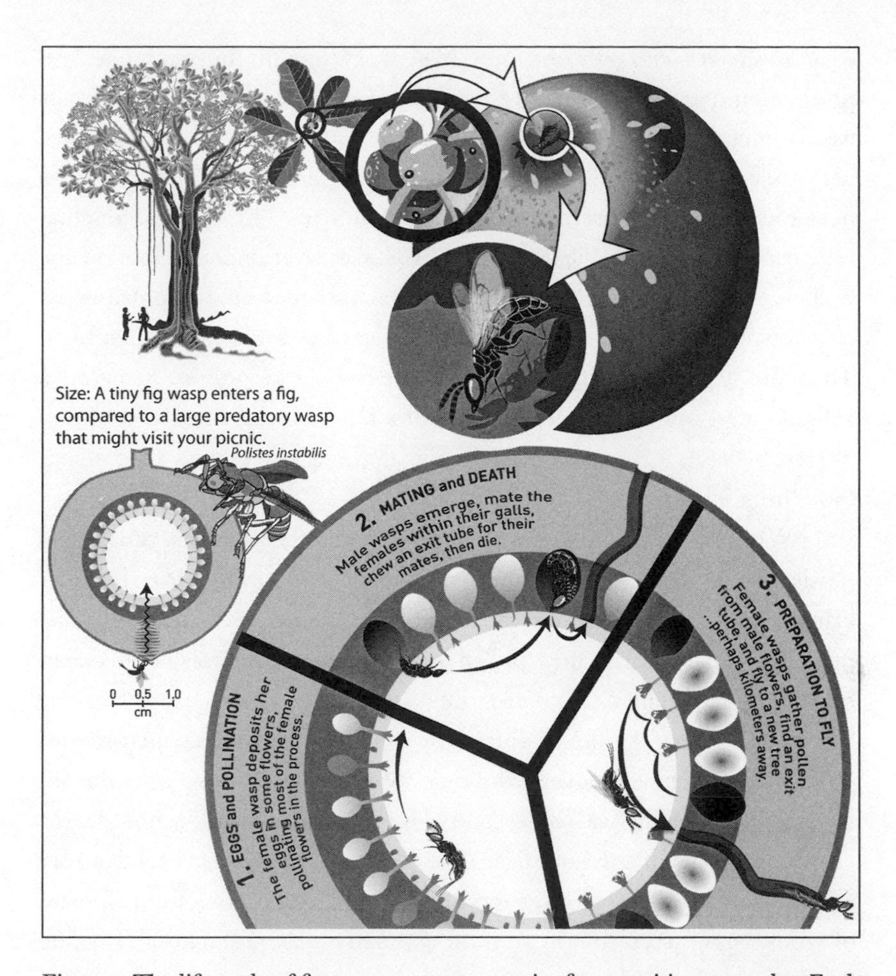

Fig. 9.5. The life cycle of fig wasps: an opportunity for parasitic nematodes. Each species of fig tree maintains its own species of live-in pollinating wasps. The minute, fertilized, pollen-bearing females (an ordinary predaceous wasp is shown for scale) crawl through a hole into a fig—now a flowerhead turned outside in, ready to pollinate—pollinate the flowers, and lay one egg each in half of them (top). Each wasp larva matures within a single fig seed (bottom). Males hatch first, fertilize the females, which have not yet hatched (eclosed), dig an exit hole through the fig wall, and die. Fertilized females dust themselves with pollen, and fly off in search of other figs to pollinate. These wasps are parasitized by nematodes: emerging wasps are entered by young nematodes that mature during the flight to the fig where their hosts lay eggs, and the nematodes emerge, mate, and produce young that will repeat the cycle. (Diagram by Damond Kyllo)

In species with only one foundress per syconium, the nematodes' reproduction depends on that of their hosts, so they must treat their hosts well to enhance their own reproduction. In "single-foundress" fig species, nematode-bearing fig wasps therefore produce as many offspring as nematode-free counterparts. In "multiple-foundress" fig species, a nematode can colonize the offspring of other wasps, so it depends less on the welfare of its own host. Since unrelated competitors might parasitize its own wasp's offspring, a nematode benefits less by caring for its own host. Therefore, in species with several foundresses per syconium, nematodes substantially reduce the reproductive output of their carriers: the common interest between wasps and parasitic nematodes is too weak to keep nematodes from injuring their hosts.

When individuals acquire symbionts from other sources, mutualism persists if cooperation is advantageous enough. Figs and their pollinators bring complementary assets to the table: figs provide food and a safe home for the young, the wasps move pollen to other fig trees. Fig wasps can cheat, however: how can fig trees control this?

Cooperation is often enforced by not feeding cheaters, just as employers keep their employees working by not paying those who do not work. Fig-pollinating wasps descend from "seed predators," whose larvae grew up in seeds of ancestral figs. Adult wasps must find new seeds to lay eggs in, and insect seed-eaters tend to specialize on seeds of one species of plant, so fig trees could use these seed eaters as pollinators. Turning the flowerhead outside in led a syconium's wasps to mate with each other, and by making enough pollen they ensured that female wasps were dusted enough with it to automatically pollinate flowers when laying eggs in a new syconium. When fruiting, each fig tree releases a unique odor that attracts only pollinating wasps of the right species. Later, some fig trees evolved whose syconia produced so little pollen that wasps had to actively gather it, enabling them to cheat their hosts by not pollinating syconia where they laid eggs (a fig flower need not be pollinated for a wasp larva to mature in its ovule). These figs compelled wasps to gather pollen by dropping syconia with unpollinated flowers, ensuring that wasps that do not pollinate do not reproduce.

Many plants of the legume family provide homes, nodules on their roots, for "nitrogen-fixing" bacteria that make ammonia from atmospheric nitrogen. The plants supply their bacteria with oxygen, and the energy the bacteria need both to fix nitrogen and to live and reproduce, in return for the ammonia the bacteria provide. Nitrogen-fixing plants use chemical signals, modified from those used to call mycorrhizal fungi and guide their growth, to call their bacteria from the soil and build nodules to house them. The plants cannot detect, let alone exclude, all "bad applicants" that use their host's carbohydrates only to multiply, sparing none for the energy-demanding process of fixing nitrogen. Plants make their bacteria fix nitrogen by reducing oxygen flow to nodules where no nitrogen is fixed. Toby Kiers, then a graduate student at the University of California at Davis, showed how soybean plants punished non-cooperative bacteria. She replaced the air in some nodules with a mixture of 80 percent argon, 20 percent oxygen. Without atmospheric nitrogen, the bacteria could make no ammonia, and their hosts responded by halving the oxygen supply of these bacteria, halving their reproduction.

Corals and zooxanthellae also bring complementary assets to the table, corals providing a safe home and nitrogenous wastes from the prey they catch, zooxanthellae providing sugars they photosynthesize (plate 9.9). Mutualism benefits both, and persists even if zooxanthellae are "called" from the surrounding water. A coral's zooxanthellae sometimes cheat, but corals expel non-performing symbionts.

Cheating is also impractical if the fates of partners are interlinked, a circumstance called partner fidelity. In Central America, swollen-thorn acacias, *Vachellia cornigera,* house and feed stinging ants, *Pseudomyrmex ferruginea,* in their hollow thorns (fig. 9.6). These ants protect them from herbivores and encroaching vines. Colonizing ants take two years to produce their first offspring, whereas unprotected acacias die within a year. To reproduce, these ants must protect their hosts: the fates of acacias and their ants are thus intertwined. Colonies of a cheater ant, *Pseudomyrmex nigripilosa,* however, reproduce two months after colonizing, long before lack of anti-herbivore defense kills their host. They colonize ant-acacias, using the food and shelter they offer, but do not protect them. Intertwined fates,

Fig. 9.6. Like many plants, swollen-thorn acacias employ ants as defenders. The ants patrol the plant, nibbling encroaching vines and scaring off or killing herbivores. The sharp thorns house the colony of ants and their larvae, and nectaries at the base of young leaves feed ants patrolling these most vulnerable parts of the plant. The ants can reproduce only two years after colonizing. Unprotected plants die in a year: to reproduce, these ants must protect their tree. (Drawing by Damond Kyllo)

ensured when ants can only reproduce after the host would have died if unprotected, is essential to preserving this mutualism.

On a reef in Jamaica a college student, Janie Wulff, noticed sponges of three very different species growing intertwined in ways that avoided smothering each other (plate 9.10). She suspected that all three species benefited from being intertwined with each other. Later work, on reefs off Panama's Caribbean coast, showed that these three species were *Iotrochota birotulata,* broken fragments of which reattach most readily to reef platforms; *Amphimedon rubens,* which suffers least when sediment accumulates, and is distasteful to predatory starfish; and *Aplysina fulva,* which is least easily broken or ripped away by storms, and least vulnerable to predatory angelfish. Sixteen years of observation and experiment on these reefs allowed Wulff, now a professor, to show that each species grew faster and survived better when intertwined with the others. Growing intertwined allows these three species to pool their contrasting powers to resist different types of hazard for their common good. If a starfish eats the base of an *Iotrochota,* those fragments stuck to a starfish-resistant *Amphimedon* will live to see another day. If a storm uproots an *Amphimedon,* fragments stuck to storm-resistant *Aplysina* will survive, and so forth. Mutualism persists because these sponges have similar lifetimes and because, once intertwined, they stay intertwined. Hurting one's partner hurts oneself, for the partners' fates are interlinked.

If one partner can survive without the other, however, things change. A parasitic sponge, *Desmapsamma anchorata,* also grows intertwined with these mutualists. It grows much faster than its hosts, is much shorter-lived, and reproduces quickly by fragmenting, so it does not depend on the long-term prosperity of its hosts. This species benefits from the useful qualities of its partners but contributes nothing in return. Instead, it smothers and kills those branches of its partners that it grows on. Its much faster paced lifestyle favors a hit-and-run strategy. Like a virulent disease, *Desmapsamma* must infect new hosts before killing its current ones. Although the mutualists can only defend themselves against the parasite by shedding infected parts, they are nonetheless far more common than their parasite.

Cheating can also be controlled by partner choice: testing potential symbionts and accepting only the most cooperative. Nocturnal bobtail squid, *Euprymna scolopes,* employ bioluminescent bacteria to make the squid's shade of darkness match that of the surrounding water. This "adjustable shading" hides the squid from both prey and predators. Within two hours of hatching, these squid attract bacteria from the surrounding water, but then they eject or kill all but their chosen species, *Vibrio fischeri.* To occupy their homes within the squid, however, the vibrios must surmount an obstacle course as severe as anything a fairytale swain faces to win his princess. Vibrios can become residents inside the squid only if they are mobile, survive acid conditions and a potent microbicide, and produce sufficient light. These squid expel 95 percent of their symbionts every morning. As the best light-producers are most securely attached to their homes, they escape these expulsions, so the squid's symbionts remain cooperative.

Unlike complex human economies, which have laws to punish cheating in cooperative enterprises of all kinds, how cheating is controlled in natural ecosystems depends on the nature of the mutualism and the characteristics of its participants. Nonetheless, mutualism is as essential to the productivity and diversity of natural ecosystems as cooperative enterprise is to the health of human economies.

Major Transitions of Evolution

The interplay of individual adaptation, social cooperation, speciation, and mutualism between species drives evolution. This interplay sometimes yielded major evolutionary transitions, where "evolution the tinker" combined and modified already existing kinds of organisms to make new ones with new abilities. Life began with the coordination of metabolic and replicatory processes in a way that caused the first living things to harness energy to replicate themselves. This made natural selection possible: it can only affect reproductive entities. Here, we consider transitions involving the combination and modification of preexisting organisms, such as an archaean's transformation of bacteria into live-in mitochondria, thereby evolving eukaryotes, and the evolution of clonal clumps of cells into complex multicellular organisms.

Many evolutionary transitions are what Haldane and Huxley called evolution by aggregation, "the joining together of a number of separate units to form a super-unit. This is often followed by a division of labour among the various units, which of course . . . [can turn] a mere aggregation into an individual." Such transitions are either egalitarian, involving cooperation among non-relatives, or fraternal, involving cooperation among relatives. The evolution of eukaryotes was egalitarian, that is to say, no partners stopped reproducing, for genes never spread by sacrificing their bearers' reproduction to help non-relatives. The evolution of metazoans and insect societies were fraternal transitions involving cooperation among relatives. Here, division of labor begins between a few reproductive cells, and many non-reproductives that spread their genes only by helping their relatives reproduce. The distinction between egalitarian and fraternal transitions reflects contrasts in the common interests of the autosomal genomes in the organisms involved.

The most decisive egalitarian transition was when archaean hosts domesticated the bacterial ancestors of mitochondria. Taming these live-in bacteria posed daunting challenges that aroused radical responses. Unlike prokaryote genes, eukaryote genes encode several translated DNA sequences, exons, separated by sequences of apparently useless DNA called introns. These introns are remnants of bacterial "selfish genetic elements" that could move from one chromosome to another. In genes shared by all eukaryotes, these introns tend to occupy the same position in all species, suggesting that they were incorporated before eukaryotes could evolve suitable defenses, let alone diversify. As the symbiosis was evolving, dying symbiotic bacteria must have released DNA, allowing its selfish genetic elements to bombard their host archaean's genome which, unlike bacterial genomes, could not control their incorporation. This bombardment posed a major problem. After messenger RNA was transcribed from a gene, these nonsense introns had to be excised by a special protein and the translatable portions joined together before a ribosome could translate the RNA into the protein this gene encoded. This bombardment was greatly reduced when membrane-programming genes acquired from bacterial symbionts formed a porous "nuclear membrane" separating the chromosome from the symbionts and their selfish genetic elements. Moreover, introns could now be removed from messenger RNA inside the membrane, while the edited RNA was translated into proteins by ribosomes outside the membrane. This membrane-surrounded chromosome was the original nucleus. Nonetheless, bombardment by selfish DNA also required DNA repair on a large scale. This was accomplished by the fusion (mating) of two archaean individuals and pairing of their chromosomes: the sound parts of each were used to repair damage in the other. Thus, what were once considered three distinct evolutionary transitions, domesticating mitochondria, evolving nuclei, and evolving orderly sexual reproduction, were aspects of one event. These developments were so improbable that successful eukaryotes evolved only once.

The precise pairing of chromosomes required for meiosis favored a "canonical genome" of a specific set of loci, identically arranged into chromosomes, for each interbreeding population. It became advantageous to choose mates from one's own population, to avoid producing inferior or inviable hybrids. It also nearly eliminated horizontal gene transfer, allowing coevolution among a population's loci.

Domestication of these symbionts resulted from their becoming dependent on their hosts. Resident symbionts benefited from excluding immigrants that might oust them or overexploit their host. Therefore, after they spread to all available hosts and excluded immigrants, they could only spread from a host to its descendants. These bacteria now depended utterly on their hosts' welfare: a symbiont's reproduction hinged on how well it and its fellows helped their host reproduce.

Sexual reproduction allows horizontal symbiont transfer if a sperm carrying symbionts fertilizes an egg with symbionts of different genotype. Kin selection causing symbionts to favor relatives over non-relatives may promote killing unrelated symbionts, creating a "civil war" that could devastate the host. Most sexual species prevent such mixing by letting zygotes inherit symbionts from only one parent, usually the mother. Passing on symbionts only through the mother creates a new danger: now, mitochondria benefit by programming all-female sex ratios. They sometimes do so. Their hosts' autosomal genes, however, share a common interest in producing equal numbers of offspring of each sex, so selection on the host's genes soon restores this equality. When there is genetic conflict between hosts and mitochondria, the hosts, whose genomes are much larger and more varied, prevail. Uniparental transmission of mitochondria (and chloroplasts) and the resulting conflict between organisms and their organelles over host sex ratio are relics of the genetic conflicts involved in domesticating these organelles.

"Fraternal" transitions included evolution, less than a billion years ago, of many-celled animals (metazoans), plants (metaphytes), and fungi (fig. 9.7). All many-celled animals, plants, and fungi descend from sexually reproducing ancestors, even though every sexually reproduced organism begins as a vulnerable single cell. Why?

Selection is more effective when it has more alternatives, more genetic variation, to choose from. To transform a group of cooperating individuals into an integrated individual of coordinated parts, genetic variation among groups must far exceed that among each group's individuals. In many-celled organisms, each fertilized egg has a unique genotype. Except for the occasional mutation (copying error), however, the cell divisions making the fertilized egg a many-celled organism duplicate its genome. If primitive many-celled aggregates also grew by successive cell divisions in sexually produced zygotes, genetic differences between these aggregates greatly exceeded those within aggregates. Sexual reproduction, clonal reproduction of each fertilized egg to become a many-celled aggregate, and excluding cells from other aggregates allowed selection among aggregates to override selection within aggregates. Selection favored aggregates with signaling systems that coordinated activities among a clump's cells, transforming these aggregates into genuine individuals with extraordinary diversity of functions among their cells. Cells of insects and vertebrates are differentiated into sensory systems, digestive systems, locomotory organs, brains, and so on.

These differences can arise because chemical gradients from the embryo's front to rear and its top to bottom make possible genes that are only expressed at specific levels of these gradients. Such genes promote only those developmental processes appropriate to their part of the embryo. They also create a modularity whereby a developmental mutation affects only that part of the embryo where it can be expressed.

Mutation rates are so low that organisms with fewer than a million

Fig. 9.7. The difference multicellularity can make: *Paramecium* (0.3 millimeters long), a multicellular rotifer (0.5 millimeters), and a dog, 400 millimeters tall at the shoulder. (Drawing by Debby Cotter Kaspari)

cells have few mutants. Animals and plants with trillions of cells, however, could evolve only if they were able to keep mutant cell lines, usually harmful or fatal to the organism, from spreading. In animals, three such defenses are maternal control of early stages of cell differentiation, cells (phagocytes) that eat potential cancer cells of abnormal form or genotype, and isolation of the "germ line" (reproductive cells) from the others, to keep cancers from spreading from parents to their young. In plants, rigid cell walls hinder the spread of cancers through the organism by metastasis, as so often happens in animals.

Genetic uniformity among their cells is essential to the evolution of complex multicellular organisms, just as suppressing reproductive conflict by allowing only one individual to reproduce enables the evolution of complex, stable insect societies. In social amoebae, once called slime molds, genetically distinct amoebae join to form a many-celled aggregate, a slug. After moving to a suitable place to reproduce, the slug transforms into a fruiting body made of spores atop a stalk. Social amoebae are plagued by cheating genotypes, whose bearers all become spores rather than non-reproductive parts of the stalk that allows the spores to disperse much farther.

Many-celled animals possess traces of the genetic conflicts that were overcome as they evolved. These traces include conflicts between metazoans and cancerous cell lines, and the very elaborate machinery by which they are suppressed. For example, complex animals keep their cells genetically uniform by specialized phagocytic cells that eat cells of abnormal genotype. Genetic conflicts between cells and their mitochondria, and between metazoans and their component cells, and the means by which they are suppressed, are footprints of the processes that transformed free-living bacteria into mitochondria and clonal clumps of cells into individual metazoans.

The evolution of mitochondria and metazoans offers another lesson. Selection has often transformed groups of preexisting organisms—each one an already tested module, into larger wholes. Natural selection did not engineer mitochondria or chloroplasts de novo: like a tinker, it modified bacteria engulfed by the host to make them. Likewise, evolution the tinker transformed clones of cells, each of which could reproduce and respond to environmental challenges, into integrated many-celled individuals. The processes of development of an adult from an egg can therefore respond appropriately to drastically changed conditions. One animal famous in evolutionary biology was a deformed goat: its front feet had been rendered useless by an adverse mutation. This goat walked upright on its hind legs. Walking upright created stresses on its bones and muscles that triggered extensive adjustments in its skeleton and musculature, enabling it to cope with the lack of front feet. These developmental responses to new conditions reflected responses in the cells of these structures, born of the tendency toward self-preservation. As Sewall Wright remarked, "structure is never inherited as such, but merely types of adaptive cell behavior which lead to particular types of structure under particular conditions."

TEN

The Last Transition

How Thought and Language Evolved

KONRAD LORENZ, ONE OF ONLY THREE students of animal behavior to share a Nobel Prize, considered the origin of life—the origin of beings that could harness energy from their surroundings to reproduce themselves—and the evolution of conceptual thought and language the two crucial turning points in the earth's history. He observed that life "came into being with the 'invention' of a structure possessing the power to acquire and retain information, and capable at the same time of amassing out . . . of the energy around it a sufficient quantity for it to keep the flame of knowledge burning." He viewed natural selection of random mutation as "the method of the genome, perpetually making experiments, matching their results against reality, and retaining what is fittest." Like a scientist seeking objective knowledge, natural selection, through "trial and error," refines a population's "hypothesis" of how to live and reproduce in its habitat. In short, genomes store a population's knowledge of how to use its surroundings to live and multiply, and natural selection improves this knowledge. This circumstance allowed natural selection to produce a succession of living beings that transformed the world.

Lorenz also saw that conceptual thought and language made it far easier to pass learned knowledge from one generation to the next, thus making it heritable. A jackdaw can communicate to another that cats are dangerous only if a cat is present. Speech abolishes this restriction. Thus language enabled cultural evolution to become the primary engine for modifying human behavior. It eventually enabled rapid change.

Abstract thought enables one to solve complex problems. Language conveys thought, enabling far more intricate cooperation. Language and conceptual thought allow wisdom learned by experience to be passed on to future generations. Five millennia ago, thinking beings invented writing. Writing made it easier to preserve and diffuse knowledge, which in turn allowed widely separated people to coordinate their activities. Later, printing and improved transport greatly amplified the increase and spread of knowledge and enabled ever more intricate coordination of cooperative enterprise on ever wider scales. Consequently, human beings now dominate this planet.

Cultural Versus Genetic Inheritance: Pros and Cons

Genomes "learn" only by allowing natural selection to test variations, mutations, in their hypothesis of how to live and reproduce in their habitat. In eukaryotes, genomes are organized to allow selection to sift beneficial from harmful alleles more easily, making evolution quicker. Sexual reproduction "levels the playing field," enabling natural selection to "judge" a new allele by its own contribution to its bearers' fitness rather than by the fitness of the genotype in which it first appears. Fair meiosis allows an allele to spread only if it benefits its bearers, thus making evolution adaptive.

Among social animals, as among small tribes, a cultural novelty spreads from parent only to offspring and near neighbors. Faults would appear soon enough to prevent this novelty's spread. Predatory animals from small bats to large cats teach their young how to tackle dangerous prey by killing prey in their young's presence, then letting them imitate these techniques on successively less injured living prey. Such killing techniques spread only if they work. As societies grew and communication and transport improved, ideas traveled farther and faster, and were combined in useful ways, so cultural evolution accelerated. New means of cultural transmission such as schools and universities made combining new ideas easier. In some schools and universities, interactions among scholars tended to ensure that new ideas were taught only if they provided better ways of explaining or exploiting their world or of improving social life. The knowledge and technology thus discovered and transmitted allowed human beings to take over the planet. These explosive changes, however, exacted a price. Technological versatility and fast long-distance communication led human activity to outrun morality. Traditional morality does not envisage all the ways new developments can hurt others—from unsustainably destructive ocean fishing techniques, and new fertilizers or drugs for birth control that poison groundwater, to banks growing "too big to fail," big enough that their failure would ruin their nation's economy. These problems take public resources that are needed for other urgent problems. Moreover, rapid long-distance communication, so useful in some ways, also spreads dangerous delusions and destructive hatreds. Speculators caused devastating financial crises by spreading the belief that stock or house prices would increase endlessly. The current epidemic of internet-driven political, ethnic, and religious hatred may cause yet more damage. Cultural evolution lacks the quality control of its genetic counterpart.

Steps Toward Conceptual Thought and Language: Knowing Objectively

Following Lorenz, we assume that "all human knowledge derives from an interaction between man as a physical entity, an active, perceiving subject, and the realities of an equally physical external world, the object of man's perception," and "everything reflected in our subjective experience is intimately bound up with, based on, and in some way identical with physical processes that can be objectively analyzed." These assumptions are our only hope for attaining a scientific understanding of thought and language. The work of the neurobiologist Jean-Pierre Changeux, seen in the light of his dialogue with the philosopher Paul Ricoeur, suggests that these assumptions bear fruit.

"Evolution the tinker" produced conceptual thought by combining many preexisting animal abilities, including knowing objectively, forming abstract concepts and theories, reasoning by analogy, and controlling one's actions well enough to learn new techniques for making tools, catching prey, processing food, and the like.

WHY DO ANIMALS NEED OBJECTIVE KNOWLEDGE, AND HOW DO THEY GET IT?

An organism must have objective knowledge of its environment to survive. As François Jacob, a French molecular biologist who shared a Nobel Prize with Jacques Monod, remarked, "No matter how an organism investigates its environment, the perception it gets must reflect . . . 'reality' and, more specifically, the aspects of its reality which are directly related to its own behavior." Moreover, "the increase in performance that accompanies evolution requires a refinement of perception, an enrichment of the information received concerning the environment." As we shall see, the need for social perception in group-living primates also favored the ability to form abstract theory and reason by analogy. Indeed, the objective knowledge natural selection encoded in some animals' genomes enabled them to learn objective truths about their surroundings. Brains provide their bearers good enough representations of the natural world to enable them to act appropriately, interpret otherwise unusable sensory impressions, and, at least in apes and human beings, subjectively simulate possible actions and events so as to predict their results and respond accordingly. How could this happen?

The first step to making sense of one's surroundings is to distinguish the relevant and register only those signals that mean something. A scientist singles out observations relevant to the hypothesis she is testing, ignoring the rest; an animal singles out signals indicating food, predators, obstacles to avoid, and other useful data from irrelevant sensations. Long before animals could learn what signals to respond to, natural selection was doing so for them. Although a still, and a moving, black speck both register on a frog's retina, retinal nerves only transmit signals of the moving speck, which usually signifies flying prey. To reproduce, a female rainforest cricket must distinguish calls of males of its species from the cacophony of other night sounds in the forest—thus solving "the cocktail party problem." Thus cricket "ears" are far more sensitive to these mating calls than to sounds of even slightly different pitch. Many moths and katydids have special organs to detect the echolocation calls of predatory bats "homing in" on them.

Similarly, like honeybees (for whom color helps to distinguish the flowers they seek) we can discern an object's color, regardless of the color and intensity of the light shining upon it, be it the bright light of midday or the reddish light of sunset. By trial and error, natural selection has long been shaping nervous systems that sift sensory inputs and integrate relevant signals into representation of their bearers' surroundings good enough to let them escape predators, find food, and reproduce.

In many invertebrates and lower vertebrates, particular stimuli trigger instinctive, inflexible, often elaborate "fixed motor patterns," in which learning plays no role. The French insect behaviorist Jean-Henri Fabre showed that a wasp that stocks its nest-cells with spiders for its larvae to eat continues to stock and build a cell if one removes the egg and spiders already placed there. Moreover, if all cells are removed after the wasp has begun to plaster its nest with an outer covering, it continues to plaster the bare surface where the nest was. Fabre gave many other examples of fixed motor patterns in insects. Lorenz described how a goose continued the movements involved in returning an egg to the nest if deprived of that egg. Splitting fixed action patterns into individually controlled subroutines allows animals to respond more flexibly to happenings around them, for which they need to understand their surroundings better.

Some animals, like cats, monkeys, and human beings, unconsciously integrate sensory inputs to abstract the presence and properties of an object from their maze of sensory impressions of it. How the brain infers this object—its size, shape, color, and the like, as the common source of views of it from various directions, in different light, at different distances—is the true miracle of vision. The mathematical physicist Hermann Weyl remarked that how a person abstracts an object's existence and properties from diverse impressions of it is analogous to how theoretical physicists abstract objects such as electrons or black holes from diverse data. The difference is that the physicist abstracts consciously, whereas our abstraction of objects from sense data is unconscious. The ability to abstract is independent of consciousness: organisms evolved this ability long before they evolved "higher mental faculties." Once an animal can distinguish objects and recognize them when encountered again, it can distinguish sets of objects sharing specific features, as a monkey classifies potential predators into flying eagles, large walking carnivores, and dangerous snakes, making a distinct alarm call for each category. The ability to resolve one's visual data into objects with specific properties and propensities also allows an animal to make much more sense of its surroundings, remember relevant aspects far more easily, and judge more reliably how its surroundings might change and how these changes might affect it. Although animals infer objects unconsciously, doing so offers them much to think about.

Brains organize sensory data in useful ways. Mobile animals need to assess the distance and direction to an object, be it prey, predator, or passageway. Large monkeys in tree crowns must be aware enough of their surroundings to know just where to jump. They must have good hand-eye coordination, which demands precise control over the hooked hands by which they swing from branch to branch. The philosopher Immanuel Kant rightly claimed that human brains innately locate objects around them in three-dimensional space—one of many innate frameworks that organize animals' sensory input. Locating objects in three-dimensional space is so deeply rooted that all human languages speak of non-spatial relationships in spatial terms—"a king's rank *is above* a duke's," "*Behind* Obama's act is the intention . . ." Did our way of visualizing spatial relationships provide an organizing template for other human thought?

Brains evolve in accord with their owners' needs and sensory capacities. Once freed by bipedal walking, hands became useful not only for making and using elaborate tools, but also in sensing objects and communicating, whether with dog, horse, or person. Among a baby's earliest essays in coordination is seeing, reaching for, touching, grasping, and feeling an object. Brains enable more skillful hand use, and hand use helps shape brain evolution—in various ways. Some geniuses at working with their hands, whether musicians, artists, or machinists, may be poor at abstract thought, whereas some abstract thinkers are clumsy with their hands.

HOW SOCIAL LIFE CAN HELP THOUGHT AND LANGUAGE EVOLVE

Social behavior in mammals favors the integration of skills needed to evolve thought and language. Animals such as wolves, meerkats, and monkeys live in groups, whose members help each other find food, avoid predators, and defend their group's feeding territory. To avoid perpetual disputes over priority of access to food or mates, members of larger groups sort themselves into dominance hierarchies, "pecking orders." A social ring-tailed lemur, worse at laboratory tests of manipulative skill than the lowliest marmoset, can recognize each of its fellow troop members and remember the hierarchical ranks and idiosyncrasies of its 11 adult fellows (plate 10.1). Their social structure, however, is far simpler than baboons'. High-ranking lemurs, unlike high-ranking baboons, do not intervene in fights to protect younger but independent relatives.

As we saw early on, a baboon in an 80-member troop recognizes each of its 79 fellows by sight or voice, and remembers its sex, and the relative dominance and degree of relatedness of each pair. Their social knowledge is syntactic: a baboon is undisturbed by recordings of a dominant threat-grunting and a subordinate screaming, but is shocked to attention by recordings of the subordinate threat-grunting and the dominant screaming. It distinguishes the actor, the act, and the one acted upon: is this distinguishing subject, verb, and object? To reproduce successfully, a baboon must know its place in the hierarchy. From outcomes of pairwise interactions, baboons infer the rank-order of all the troop's members. This is an abstract, objective, economical theory summarizing knowledge of who outranks whom, which a baboon must have to succeed in life. When new members enter the group, baboons adjust their theory appropriately. Similarly, they infer their troop's matrilineal kinship network from interactions, past and present, among its members. It appears that natural selection has organized the minds of baboons to predispose them to identify "patterns and rules underlying other baboons' behavior."

Children learn to talk with astounding speed. They abstract syllables and words from the speech they hear, and learn how to organize their words into sentences. Noam Chomsky showed that children have an innate ability to distinguish a subject—a noun, a noun phrase including adjectives and adverbs, or a whole subordinate sentence serving as a noun—from a verb, or an object of the verb's action. Their order differs in different languages, but these distinctions provide an innate grammar, programmed in their brains, by which children organize words into sentences. Deaf children have developed sign languages organized by the same innate grammar as spoken languages. The same brain lesions that interfere with speech also interfere with sign language. This innate grammar of communication, geared to linking cause and effect, appears to organize our thought as well as our speech: it must have evolved long before speech.

Anthropologists have taught several apes to communicate with them by signs (apes cannot speak words). An ape can learn signs for classes of objects, and for a few actions. Sometimes, they distinguish subject and object by sign order. They like using signs to communicate with each other. They lack an innate grammar, however, so they learn slowly, like people learning how to write. Human beings may have used sign language before they could speak: even now, people who cannot understand each other's speech try to communicate, often successfully, by signs, gestures, and body language.

Our brains are organized for social perception. We recognize our fellows. We interpret their emotions from facial expression and tone of voice. As Darwin knew, a human facial expression normally represents the same emotion the world around. Like baboons and chimpanzees, we quickly learn kinship, who belongs to our group, and our own dominance relative to those we deal with. Do our patterns of thought and speech reflect the demands of social life? Social life gives us plenty to think and talk about.

In sum, objective inference, reasoning by analogy, and inferring (however unconsciously) theoretical constructs that economically summarize social knowledge all enhance the fitness of some non-human animals.

THE MYSTERY OF CONSCIOUSNESS

Consciousness is the realm of our subjective experience, our self-awareness. As Hermann Weyl remarked, "This inner awareness of myself is the basis for the understanding of my fellow man. . . . Even if I do not know of their consciousness in the same manner as of my own, nevertheless my 'interpretative' understanding of it is . . . of indisputable adequacy. Its illuminating light . . . also reaches, though with ever increasing dimness and uncertainty, deep into the animal kingdom."

The molecular biologist Jacques Monod argued that consciousness includes a capacity for simulating potential actions and their consequences. We, like chimpanzees, use it to predict another's behavior by imagining what we would do were we in her place. Although conscious, baboons cannot do this. Apes use their consciousness as simulator to solve problems by simulating—imagining—the consequences of potential solutions, whereas monkeys and raccoons act out one solution after another. Lorenz tells of an orangutan placed in a room with a box at one end and a banana too high to reach at the other. He sat frustrated until he looked at the box, then at the banana, imagined the box as under the banana, and promptly moved it to make this so. He apparently simulated the effect of moving the box under the banana before moving it. Dreaming presupposes imagination—a capacity to simulate. Darwin knew that dogs, cats, and horses dream.

Because consciousness is subjective, some biologists ignored it as inapproachable by objective science. Consciousness, however, is an integral part of the toolkit natural selection endowed us with, by which we know and understand those around us. Consciousness underlies "theory of mind," by which we infer how another feels and thinks by analogy with how we would feel if we were in that person's shoes. This ability is essential to human social life: autists, who lack this ability, are socially handicapped. Reasoning by analogy is also an essential feature of human thought.

To be more precise, a student of Changeux, Stanislas Dehaene, defined consciousness as "conscious access." A person has "conscious access" to a stimulus if he is aware he perceived it. Dehaene found that images people do not consciously see may affect some parts of their brains, and even influence their behavior, but conscious perception is shared throughout the cortex. This last feature lets researchers detect conscious perception in animals without speech. This approach shows great promise.

HOW PLAY CAN BRING FORTH OBJECTIVE KNOWLEDGE

Curiosity, play, and exploratory behavior develop the sensory, social, and motor skills, and the coordination among them, that animals need to live. Apes watch their hands when manipulating objects: could their curiosity lead them to view their hands, and themselves, as objects like others they see? Such an insight could prompt dawning self-awareness, and promote ability to understand others by analogy with their own feelings.

Curiosity and exploratory play also favored conceptual thought. Lorenz saw exploratory play as a disinterested, objective interaction with the world. An animal tries out many behaviors to investigate an object with no apparent use, or explores a new home range without seeking to hide or find food. Adult primates learn new things from the exploratory play of their young, just as we of an older generation learned new things about computers from our children. Exploratory play and learning are for their own sake, not for achieving a specific goal. Lorenz remarked: "As a result of this apparently small step forward, there emerges an entirely new cognitive process which is in essence identical to human investigation, and which leads without an essential break to . . . scientific research. The connection between play and investigation . . . is still fully preserved in adult human investigation, whereas in adult animals it tends to disappear." Indeed, in our brains reason and emotion are inextricably intertwined: they cannot really be divorced. The best science is obviously driven by the anticipated pleasure of new and beautiful discoveries.

The Dutch historian Johan Huizinga argued in his book *Homo Ludens* (Playing Man) that play was a decisive influence on the shape of all human culture, and on human ability to think and learn. In social animals, play improves social skills by ritually enacting social behavior. Among dogs, play begins with a ritualized invitation, the "play bow." It entails rigid rules against injuring playmates—"you shall not bite, or bite hard, your brother's ear." Play can represent something, as when children dress up to play the role of other beings. Play as representation gave rise to art and science. Play can be a ritualized contest for something, as the elaborate dances of male manakins (which may have been learned by playing) are a contest to attract mates. Play as ritualized contest gave rise to phenomena as varied as Greek drama, knightly tournaments, law courts, and football. Yet the original motive for play is the enjoyment, the fun, of playing. Dogs and children, like those who first played football, chess, or music, play for the fun of it. Talented mathematicians and scientists were and still are attracted to their subjects in a childlike spirit of play, for the "pleasure of finding things out." Such play can create genuine beauty, as in poetry, music, drama, mathematics, or Newton's law of gravitation.

Mathematics is a form of ritualized play, a game where the prize goes to whoever deduces the most intriguing and beautiful patterns from a well-chosen set of axioms (basic assumptions and concepts). Whole numbers reflect our workaday need to count (many vertebrates can count to three), but Euclid's proof that there are infinitely many prime numbers (numbers such as 2, 3, 5, 7, 11, 13, 17, . . . which are divisible only by 1 and themselves) was a form of play. Later mathematicians proved that the ratio of $n/(\ln n)$ to the number of prime numbers less than or equal to n became ever closer to 1 as n increased. There are 168 prime numbers smaller than 1,000 and 50,847,478 smaller than 1,000,000,000, compared with the predicted 145 and 48,254,942. These discoveries from play, however, are immaterial, universally valid truths. Our minds may be made of matter, but in play we can discover immaterial, universal truth. This is not the least of the wonders of nature.

The convergence between mathematical and artistic play is revealed in the symmetries of one-dimensional friezes, one-dimensional repeated patterns, and two-dimensional wallpapers. A symmetry is an operation—a translation, rotation about a point, or reflection in a line, that leaves the frieze unchanged (maps it onto itself). The simplest frieze's one symmetry is translation over one or more units:

↗ ↗ ↗ ↗ ↗ ↗ ↗ ↗ ↗ ↗ or ¶ ¶ ¶ ¶ ¶ ¶ ¶ ¶ ¶ ¶.

A second type of frieze is illustrated below, where the repeating unit is ↗ ↘ or ╖ ╜ :

➚ ➘ ➚ ➘ ➚ ➘ ➚ ➘ ➚ ➘ or ╖ ╜ ╖ ╜ ╖ ╜ ╖ ╜ ╖ ╜

These are also symmetric under glide reflection (translation by a half-unit, followed by a reflection in a horizontal line, which on the left passes through the midpoints of the arrow shafts and on the right joins the horizontal parts of the symbols. A third type of frieze,

⇨ ⇨ ⇨ ⇨ ⇨ ⇨ ⇨ ⇨ ⇨ ⇨

is symmetric under translation and under reflection in the line bisecting the arrow shafts. It can be proven that only four other types of frieze symmetry exist:

4. ∧ ∧ ∧ ∧ ∧ ∧ ∧ ∧ ∧ ∧ ∧ 5. ∧ ∨ ∧ ∨ ∧ ∨ ∧ ∨ ∧ ∨
6. ✧ ✧ ✧ ✧ ✧ ✧ ✧ ✧ ✧ 7. # # # # # # # # #

The fourth is symmetric under reflections in vertical lines bisecting the figures or passing between them; the fifth under glide reflections, reflections in lines bisecting the figures, and 180-degree rotations about axes perpendicular to the paper, passing through points along a horizontal line bisecting the wedges, midway between successive wedges; the sixth under reflection in the horizontal line bisecting all the figures, or in vertical lines bisecting or passing between the figures, or 180-degree rotations around the figure centers or points between the figures; and the seventh under 180-degree rotations about axes perpendicular to the middle of a symbol or midway between two symbols. Mathematicians playing in the meadow where art and mathematics join proved that there are seventeen types of wallpaper symmetry, which Arab artists discovered empirically (fig. 10.1).

For animals, play is a powerful generator of useful knowledge. For human beings, play can yield yet more. Mathematical play has often developed tools vital for expressing the findings of physical science. The mathematics that allowed Einstein to express the law of gravitation in terms of how stars and planets curve space and how the curving of space shapes their motions also predicted never-imagined phenomena such as black

Fig. 10.1. Mathematics as art: two forms of play combined. This diagram represents one of seventeen types of wallpaper symmetry discovered empirically by Muslim artists decorating the walls of mosques, and first demonstrated theoretically in 1924 by the mathematician George Polya. Here, the pattern is unchanged by reflection in straight lines such as the four intersecting lines shown in the unit at top left, 90-degree rotations about vertical axes passing through points where four lines cross, and 180-degree rotations about a vertical axis through points where two lines meet, such as the circled dot, or by translation across one unit, either up, down, or sideways. (Diagram by Mary Bruce Leigh)

holes and gravitational waves. Such facts impelled the Nobel Prize–winner Eugene Wigner to wonder just why mathematics have proved "unreasonably effective" in physics. Play is often associated with beauty. The grandfather of modern ecology, G. Evelyn Hutchinson, remarked that education should "aim at seeing life as a sacred dance or as a game in which the champions are those who give most beauty, truth and love to the other players." Like the French mystic Simone Weil, Hutchinson thought the true definition of science is that it is the study of the beauty of the world.

ELEVEN

What Have We Learned, and What Is Still Unknown?

What We Think We Know

Nearly four billion years ago, alkaline, hydrogen-rich water emerging through deep-sea vents from the magma below met more acidic ocean water rich in carbon dioxide. Did the thermodynamic disequilibrium between emerging vent water and ocean-bottom water subsidize successive stages in the origin of living beings that deployed energy from this disequilibrium to reproduce themselves? If so, processes without purpose led to beings organized to live and multiply, an event philosophers consider paradoxical. The reproduction of early organisms must have hinged on molecules that encoded their "hypothesis of how to live and reproduce," how to harness and deploy energy and matter to do so. Natural selection improved their descendants' hypothesis, so they multiplied and transformed the earth, making it more hospitable to life. Competition for the resources needed to reproduce favored diverse ways of life, leading to interdependence among forms with complementary methods of procuring resources. Adaptive diversification and cooperation within and among species enabled the evolution of diverse, productive ecosystems where competition was intense. Relationships of interdependence became ever more intricate and extensive.

To understand how adaptation evolves requires knowing how offspring inherit their parents' characteristics. These characteristics are transmitted from parents to their young by genes, molecules that collectively encode the parents' hypothesis of how to live and reproduce—the basic processes governing an organism's development and behavior. For selection to spread alleles that cause adaptive evolution, gene replication must be organized and coordinated. The elaborate process of meiosis, assigning genes from a diploid cell to haploid gametes, makes possible an orderly sexual reproduction with extensive exchange of genes among different genotypes. By exchanging genes among homologous chromosomes, sexual reproduction lets natural selection test an allele by its own contribution to fitness rather than by the fitness of the genotype where it first appeared and to which, were it not for sexual reproduction, it would still be bound. This property allowed selection to distinguish far more effectively between useful and harmful alleles, and favored alleles that were "good mixers," enhancing the fitness of all genotypes they occurred in. Thus sexual reproduction favored a modular organization of the processes through which genes program an organism's growth and behavior, so selection could change one feature without altering others. Finally, orderly sexual reproduction with recombination between homologous chromosomes created a common interest among an organism's genes in fair meiosis, which allowed alleles to spread only if they benefited their carriers, adapting them to their environments.

These rules, which leveled the playing field for competition among alleles and ensured that alleles spread only if they benefited their carriers, laid the foundation for adaptive evolution. Thus natural selection could favor social groups that later evolved into individuals in their own right, as in the evolution of multicellular organisms and closely integrated honeybee colonies. Trade-offs between different ways of life could favor splitting a population into a different species for each. Finally, where two species benefited by pooling contrasting abilities for the common good, natural selection favored mutualism. Mutualism was crucial both to the evolution of eukaryotes and to the function of ecosystems. Together, adaptation, diversification, the evolution of group life, and the evolution of mutualism account for the evolution of complex organisms living in complex, interdependent communities.

Frontiers of Gross Ignorance

The first frontier is the origin of life. How did living beings organized to survive and reproduce arise from lifeless matter? How did organisms that organize matter to further their survival and reproduction arise from processes as lacking in purpose as falling water or forest fires? We suggested how this might have happened, and provided evidence for these ideas, but a convincing, detailed account of how this happened would be as important for philosophy as for biology.

How technological capacities of organisms—ranging from the means plants use to cope with seasonal drought or intense herbivory to the capacities that enable a bird to migrate to the right place at the right time—affect the evolution of ecosystems is an urgent question. We need to learn how the degree and scale of interdependence among different species or regions hinges on the many and varied technological capacities of organisms ranging from soil microbes to bats, honeybees, and migrant whales.

A third question is, how do the interactions between species, which often involve competition and predation, shape luxuriant, diverse communities? Why are communities that human settlers first encountered—kelp beds with sea otters (rather than barrens sea urchins create when otters are killed off), and Aleutian islands with fertile grassland and no foxes (compared to islands with introduced foxes that eat the birds that fertilize the grassland, creating an unproductive tundra)—so much more productive and diverse?

The final frontier is conceptual thought and language, the achievement of the latest evolutionary transition. This achievement and the origin of life are the two biggest revolutions in the earth's history. Are our brains thinking matter? If so, how is this matter arranged and organized to allow thought? How can material brains appreciate beauty? Vision, thinking, and appreciating beauty are all reflected by activities of neurons in the brain, so thought, speech, and appreciation involve matter. What D'Arcy Thompson said of physiology in his masterpiece *On Growth and Form* applies here:

> Meanwhile, the appropriate and legitimate postulate of the physicist in approaching the physical problems of the living

> body is that with these phenomena no alien influence interferes. But this postulate, although it is certainly legitimate, and though it is the proper and necessary prelude to scientific enquiry, may some day be proven untrue, and its disproof will not be to the physicist's confusion, but will come as his reward.

The most mysterious aspect of mind is consciousness. It is basic to the experience of human beings and who knows how many other animals. Consciousness, and its latest-evolved concomitant, self-awareness, allows us to understand other people and deal with the social problems of life. As Hermann Weyl realized, this understanding "reaches, although with ever-increasing dimness and incertitude, deep into the animal kingdom." Because consciousness seemed so hard to understand, some psychologists and animal behaviorists rejected it as both impermeable and irrelevant to science—a disgraceful abdication that hindered biological understanding. The molecular biologist Jacques Monod knew better. In 1970 he thought the divide between objective knowledge and the subjective self-understanding of consciousness was still impassible, but he refused to ignore what our consciousness could tell us. Now, Stanislas Dehaene's focus on "conscious access," conscious awareness of perception, as a measurable quantity relevant to consciousness, is opening up new paths that will allow us to understand much more about both the nature of consciousness and how the brain works.

Bibliographic Essay

Many books seek to explain evolution to lay readers. Carl Zimmer, *Evolution: The Triumph of an Idea,* Harper Perennial (2006) and his well-illustrated *The Tangled Bank: An Introduction to Evolution,* Roberts & Co. (2010); and David Sloan Wilson, *Evolution for Everyone,* Delacorte Press (2007) explain the evidence for evolution and show how natural selection adapts populations to their environments. Mark Kirschner and John Gerhart, *The Plausibility of Life,* Yale University Press (2005) and Sean Carroll, *The Making of the Fittest,* W. W. Norton (2006), show how natural selection of "random" mutations—copy errors in the genetic instructions—can yield such intricate and marvelous adaptation.

Books reviewing the major achievements of evolution from the origin of life to the origin of mind include Jacques Monod, *Chance and Necessity,* Alfred A. Knopf (1971); Konrad Lorenz, *Behind the Mirror,* Methuen & Co. (1977); John Maynard Smith and Eörs Szathmáry, *The Origins of Life,* Oxford University Press (1999), and N. Lane, *Life Ascending: The Ten Great Inventions of Evolution,* W. W. Norton (2009) and *The Vital Question: Energy, Evolution and the Origin of Complex Life,* W. W. Norton (2015). Monod and Lorenz, who grew up in an age when scientists were better grounded in the humanities (Lorenz was the last to hold the professorship of philosophy at the University of Königsberg once occupied by Immanuel Kant), speak with the most insight about the evolution of mind; Lane does much the best with that fast-moving field, the origin of life; Maynard Smith and Szathmáry are clearest on drivers of major transitions of evolution.

In many ways, the most appealing, best balanced book on evolution is the first edition of Charles Darwin's *On the Origin of Species,* John Murray, London (1859), since reprinted many times. Originally, Darwin was convinced of the near literal truth of the Bible's first two chapters. Given his ignorance of how heredity worked, Darwin made an astonishingly strong case for evolution by natural selection, an extraordinary synthesis of many fields of study. The first volume of Darwin's *The Descent of Man and Selection in Relation to Sex,* John Murray, London (1871) is full of good sense and extraordinary insight about the evolution of sociality and morality.

The quotation on page 3 in my Introduction is from Hermann Weyl, *Classical Groups,* 2nd ed., Princeton University Press (1946).

Chapter 3. Adaptation, Individual and Social

Egbert Leigh discusses Barro Colorado Island's common fruit bat, *Artibeus jamaicensis,* in *Tropical Forest Ecology: A View from Barro Colorado Island,* Oxford University Press (1999), pp. 31–32.

INDIVIDUAL ADAPTATION

How to define, and discern, adaptation is a contentious subject, on which enough words have left enough lips to tempt wholesale use of Maxwell's silver hammer. Problems in defining and demonstrating adaptation are discussed by papers edited by S. H. Orzack and E. Sober in *Adaptationism and Optimality,* Cambridge University Press (2001). I outline my own views on the subject on pp. 358–387 of that book. Fisher's definition of adaptation is quoted from Sir Ronald Fisher, *The Genetical Theory of Natural Selection,* Oxford University Press (1930), p. 38. This book set forth the most coherent program of its day for developing a meaningful, testable theory of evolution by natural selection. It is full of mathematics, difficult reading even for professionals, but understanding this book is enormously rewarding.

My summary of Aristotle's views is based on two quotations from Jonathan Barnes, ed., *The Complete Works of Aristotle,* vol. 1, Princeton University Press (1984). "But since it is right to call things after the ends they realize, and the end of this soul is to generate another being like that in which it is, the first soul ought to be named the reproductive soul" (On the Soul 416b23–25). "Now mistakes occur even in the operation of art: the literate man makes a mistake in writing and the doctor pours out the wrong dose. Hence clearly mistakes are possible in the operations of nature also. If then in art there are cases in which what is rightly produced serves a purpose, and if where mistakes occur there was a purpose in what was attempted only it was not attained, so must it also be with natural products, and monstrosities will be failures in the purposive effort" (Physics 199 a34–b4). For Aristotle, the soul is an organization to function, not an individual thing: H. Putnam, *Words and Life,* Harvard University Press (1994), p. 3.

Michael Robinson discusses methods of mimicry in "Defenses against visually hunting predators," *Evolutionary Biology,* vol. 3, pp. 225–259 (1969). He describes his experiments with an insect-eating monkey in "Insect anti-predator adaptations and the behavior of predatory primates," *Actas IV Congreso Latinoamericano Zoologia Caracas,* vol. 2, pp. 811–836 (1970). Christian Ziegler and Egbert Leigh discuss mimicry in *A Magic Web,* Oxford University Press (2002), pp. 174–191. Robert Srygley describes mimicry in *Heliconius* butterflies in "Locomotor mimicry in *Heliconius* butterflies: Contrast analyses of flight morphology and kinematics," *Philosophical Transactions of the Royal Society of London,* series B, vol. 354, pp. 203–214 (1999).

Cynthia Sagers and Phyllis Coley discuss anti-herbivore defense in *Psychotria* in "Benefits and costs of defense in a Neotropical shrub," *Ecology,* vol. 76, pp. 1835–1843 (1995). Steven Vogel wrote a charming book about our heart and blood vessels, *Vital Circuits: On Pumps, Pipes and the Workings of Circulatory Systems,* Oxford University Press (1992). Cecil Murray first derived his law in "The physiological principle of minimum work I. The vascular system and the cost of blood volume," *Proceedings of the National Academy of Sciences, USA,* vol. 12, pp. 207–214 (1926). A related paper is Cecil Murray, "The physiological principle of minimum work applied to the angle of branching of arteries," *Journal of General Physiology,* vol. 9, pp. 835–841 (1926).

To derive Murray's Law, consider fluid flow in a circular pipe of inner radius R and length L, driven by a pressure, force per unit area, P applied at one end. The fluid's movement expresses a balance between this pressure and friction of the fluid against the pipe's walls. For fluid moving slowly through a narrow pipe, flow is laminar: that is, the fluid moves like a nested series of concentric hollow cylinders. The fluid in the central "rod" moves fastest, the fluid in the outermost cylinder by the wall is still. The velocity $v(r)$ of a cylinder of radius r and thickness dr is set by the balance between the force $2\pi rdrP$ on it at one end (the pressure P times the cross-sectional area $2\pi rdr$ on which this pressure acts), and friction with the slower cylinder of radius $r + dr$ next outside and the faster cylinder of radius $r - dr$ next inside. Friction with the cylinder next outside is the viscosity (stickiness) μ of the fluid, times the velocity gradient dv/dr at $r + dr/2$, $dv/dr|_{r+dr/2}$, times the area of the surface separating the cylinders, $2\pi(r + dr/2)L$. Friction with the cylinder next inside is $2\pi(r + dr/2)\mu L\, dv/dr|_{r+dr/2}$, exerted in the opposite direction. The equation expressing this balance is

$$
\begin{aligned}
2\pi rPdr &= -2\pi L\mu\,[(r + dr/2)dv/dr\,|_{r+dr/2} - (r - dr/2)dv/dr\,|_{r-dr/2}] \\
&= -2\pi Lud\,(r\,dv/dr) \\
rP + \mu L\, d/dr\,(r\,dv/dr) &= 0
\end{aligned}
$$

Its solution is $v(r) = (R^2 - r^2)P/4\mu L$. At the center, $r = 0$ and $v = PR^2/4\mu L$; along the wall, $r = R$ and $v = 0$. The average velocity V of the fluid is $PR^2/8\mu L$; thus $v(r) = 2V(1 - r^2/R^2)$. The average volume Q moving through the pipe per unit time is the average velocity $PR^2/8\mu L$ times the cross-sectional area πR^2 of the pipe, so $Q = \pi PR^4/8\mu L$, and the pressure needed to drive fluid through this pipe at a rate Q is therefore $P = 8Q\mu L/\pi R^4$. In *Life in Moving Fluids,* Princeton University Press

(1994), pp. 290–293, S. Vogel provides a clear derivation of this equation. We follow Murray and seek the value of the pipe's inner radius R that minimizes the combined cost of maintaining the volume of fluid in the pipe and moving it through the pipe, which is

$$b\pi R^2 L + PQ = b\pi R^2 L + 8Q^2 \mu L/\pi R^4,$$

where $\pi R^2 L$ is the volume of fluid in the pipe and b is the power required to maintain a unit volume of this fluid. Increasing the pipe's radius by an amount dR to $R + dR$ changes this cost by an amount $[2\pi bRL - 32Q^2\mu L/\pi R^5]dR$. The lower cost of moving this fluid just balances the increased cost of maintaining it when

$$2\pi bRL = 32Q^2\mu L/\pi R^5,\ R^6 = 16Q^2\mu/\pi^2 b,\ R^3 = 4Q/\pi\sqrt{(b/\mu)},$$

that is to say, when the cube of the pipe's radius R is proportional to the rate Q at which fluid moves through it. Since this fluid's average velocity $V = Q/\pi R^2$, $Q = \pi R^2 V$. Thus, when the pipe's radius R is optimum, $R^3 = 2\pi R^2 V/\pi\sqrt{(b/2\mu)}$, $R = 4V\sqrt{(\mu/b)}$, and the pipe's radius R is proportional to the average velocity V of the fluid moving through it. Friction of the water with the pipe wall is $\mu dv/dr|_{r=R}$.

$$\mu dv/dr|r = R = 2\mu V d(1 - r^2/R^2)/dr = -4\mu V/R$$

Since $R = 4V\sqrt{(\mu/b)}$, $\mu dv/dr|_{r=R} = -4\mu V/R = \sqrt{(\mu b)}$. When blood vessels are adjusted to minimize the cost of moving and maintaining the blood, friction with the vessel wall is the same for blood vessels of all sizes.

Michael LeBarbera compared circulatory systems—fluid transport systems—of different animals in "Principles of design of fluid transport systems in zoology," *Science,* vol. 249, pp. 992–1000 (1990). This paper tests Murray's law for both blood vessels in vertebrates and the systems of minute canals by which seawater moves through sponges. Papers by G. P. Bidder, "The relation of the form of a sponge to its currents," *Quarterly Journal of Microscopical Science,* vol. 67, pp. 293–323 (1923) and Henry Reiswig, "The aquiferous systems of three marine Demospongiae," *Journal of Morphology,* vol. 145, pp. 493–502 (1975) describe the transport systems of selected sponges. Sponges have about 10,000 flagellar chambers per cubic millimeter of sponge wall, and these flagella may account for 30 percent of the sponge's energy budget: see S. P. Leys et al., "The sponge pump: The role of current induced flow in the design of the sponge body plan," *PLoS One,* vol. 6 (issue 12), article e27787 (2011), table 2 and p. 2. Reiswig calculates the feeding rate of some tropical sponges, and the volume of water a Jamaican reef's sponges filter per day, in "Water transport, respiration and energetics of three tropical marine sponges," *Journal of Experimental Marine Biology and Ecology,* vol. 14, pp. 231–249 (1974).

Karen Warkentin describes the response of frog eggs to predation in "Adaptive plasticity in hatching age: A response to predator risk trade-offs," *Proceedings of the National Academy of Sciences, USA,* vol. 92, pp. 3507–3510 (1995).

SOCIAL ADAPTATION: WHY AND HOW DO GROUP MEMBERS COOPERATE?

Christian Ziegler and Egbert Leigh outline the problems and benefits of group life in *A Magic Web,* Oxford University Press (2002), pp. 225–230.

How Do Honeybee Colonies Surmount the Problems of Social Life?

T. D. Seeley details many aspects of honeybee behavior in *The Wisdom of the Hive,* Harvard University Press (1995). G. E. Robinson et al. describe what governs the normal sequence of tasks in a worker honeybee's life and how other workers adjust when the hive lacks workers of certain age groups in "Hormonal and genetic control of behavioral integration in honeybee colonies," *Science,* vol. 246, pp. 109–112 (1989) and "Colony integration in honeybees: Mechanisms of behavioral reversion," *Ethology,* vol. 90, pp. 336–348 (1992). Heather Mattila and T. D. Seeley show that genetic diversity is needed for an appropriate distribution of workers over a colony's tasks in "Genetic diversity in honey bee colonies enhances productivity and fitness," *Science,* vol. 317, pp. 362–364 (2007). Karl von Frisch discusses the waggle dance and other aspects of honeybee natural history in *The Dancing Bees,* Harcourt, Brace & World (1954). T. D. Seeley et al. show how bees tell other workers about the profitability of a food source in "Dancing bees tune both duration and rate of waggle-run production in relation to nectar-source profitability," *Journal of Comparative Physiology A,* vol. 186, pp. 813–819 (2000). T. D. Seeley shows how a new swarm chooses its nest site in *Honeybee Democracy,* Princeton University Press (2010).

Baboon Social Life: What They Know Versus What They Say

This section is based on D. L. Cheney and R. M. Seyfarth, *Baboon Metaphysics: The Evolution of a Social Mind,* University of Chicago Press (2007), pp. 199–283.

Challenges of Social Life: Keeping Disease at Bay

B. L. Thorne and J. F. A. Traniello discuss how ant and termite societies resist disease in "Comparative social biology of basal taxa of ants and termites," *Annual Review of Entomology,* vol. 48, pp. 283–306 (2003). B. Hölldobler and E. O. Wilson discuss leaf-cutter ants, their public health measures, and their waste management practices in *The Superorganism,* W. W. Norton (2009), pp. 407–468. Hermógenes Fernández-Marin et al. describe later work in "Nest-founding in *Acromyrmex octospinosus* (Hymenoptera, Formicidae, Attini): Demography and putative prophylactic behaviors," *Insectes Sociaux,* vol. 50, pp. 304–308 (2003), and in "Reduced biological control and enhanced chemical pest management in the evolution of fungus-growing in ants," *Proceedings of the Royal Society,* series B, vol. 276, pp. 2263–2269 (2009).

Chapter 4. Life's Common Ancestry, and Its Origin

EVIDENCE FOR EVOLUTIONARY DIVERGENCE FROM SHARED ANCESTORS

Darwin's Evidence for Shared Ancestry

Darwin gives his evidence in *The Origin of Species.* "Missing links" between lobe-finned fish and amphibians are described in Edward Daeschler et al., "A Devonian tetrapod-like fish and the evolution of the tetrapod body plan," *Nature,* vol. 440, pp. 757–763 (2006); Neil Shubin et al., "The pectoral fin of *Tiktaalik roseae* and the origin of the tetrapod limb," *Nature,* vol. 440, pp. 764–771 (2006) and "Pelvic girdle and fin of *Tiktaalik roseae,*" *Proceedings of the National Academy of Sciences, USA,* vol. 111, pp. 893–899 (2014); and C. Zimmer, *The Tangled Bank,* Roberts & Co. (2010), pp. 64–68. The land these tetrapods emerged from was humid tropical or subtropical forest: G. J. Retallack, "Woodland hypothesis for Devonian tetrapod evolution," *Journal of Geology,* vol. 119, pp. 235–258 (2011). The "missing links" between whales and land mammals are described in Philip Gingerich et al., "Origin of whales from early artiodactyls: Hands and feet of Eocene Protocetidae from Pakistan," *Science,* vol. 293, pp. 2239–2242 (2001) and in J. G. M. Thewissen et al., "From land to water: The origin of whales, dolphins and porpoises," *Evolution: Education and Outreach,* vol. 2, pp. 272–288 (2009). Jerry Coyne discusses what embryos reveal about life's common ancestry in *Why Evolution Is True,* Viking Penguin (2009), pp. 73–80.

Charles Darwin remarked that plants and animals of oceanic islands are most related to those of the nearest mainland in *On the Origin of Species,* John Murray

(1859), pp. 397–399. Sherwin Carlquist discusses characteristics favoring dispersal to distant islands in *Island Life,* Natural History Press (1964), pp. 9–84 and *Hawaii: A Natural History,* Pacific Tropical Botanical Garden (1980), pp. 81–110.

Biochemistry, Molecular Biology, and Life's Common Ancestry

The quotation is from James Watson, *The Molecular Biology of the Gene,* 3rd edition, Benjamin/Cummings (1976). How genes convey their instructions, and how the activities of their enzymes are coordinated, are described in Jacques Monod, *Chance and Necessity,* and the 4th edition (1987) of Watson's book (see especially chapters from p. 621 on).

HOW LIFE BEGAN

Following W. Martin and M. J. Russell, "On the origin of biochemistry at an alkaline thermal vent," *Philosophical Transactions of the Royal Society,* series B, vol. 362, pp. 1887–1925 (2007), Nick Lane suggests how life began in *Life Ascending: Ten Great Inventions of Evolution,* Norton (2009), pp. 8–59 and *The Vital Question: Energy, Evolution and the Origins of Complex Life,* Norton (2015). The reaction under the sea bottom of olivine, $(Mg_3Fe)(SiO_4)_2$, with water, H_2O, produces the hydrogen in vented fluids as well as serpentine, $Mg_3Si_2O_5(OH)_4$, and magnetite, Fe_3O_4. This reaction could be schematized as follows:

$$6[(Mg_3Fe)(SiO_4)_2] + 14H_2O \rightarrow 6[Mg_3Si_2O_5(OH)_4] + 2Fe_3O_4 + 2H_2.$$

Abiotic reactions in alkaline vents produce methane from H_2 and CO_2, and amino acids from H_2, CO_2, and ammonia, NH_3: J. P. Amend and E. L. Shock, "Energetics of amino acid synthesis in hydrothermal ecosystems," *Science,* vol. 281, pp. 1659–1662 (1998), and B. Ménez et al., "Abiotic synthesis of amino acids in the recesses of the oceanic lithosphere," *Nature,* vol. 564, pp. 59–63 (2018). In bacteria, the overall result of the acetyl coenzyme A pathway is to combine hydrogen, H_2, with carbon dioxide, CO_2, to form vinegar (CH_3COOH) and water by the formula $4H_2 + 2CO_2 \rightarrow CH_3COOH + 2H_2O$. In methane-generating archaea (methanogens), the same pathway releases energy by combining hydrogen with carbon dioxide to make methane (CH_4) and water by the formula $4H_2 + CO_2 \rightarrow CH_4 + 2H_2O$.

Chapter 5. Diversification

COMPETITION, INNOVATION, AND DIVERSIFICATION

On pp. 246–291 of *Nature: An Economic History,* Princeton University Press (2004), G. J. Vermeij shows how innovation and diversification allowed organisms to transform the earth into a suitable habitat for life. The quotation from Georges Bernanos, translated by Egbert Leigh, is taken from the midpoint of his essay "La liberté pour quoi faire," in his book of that title, published by Éditions Gallimard (1953).

Early Microbial Life: The Chemical Achievement

The evolution of metabolic pathways that help to make the earth a suitable home for life are described in A. G. Fischer, "Biological innovations in the sedimentary record," pp. 145–157 in H. D. Holland and A. F. Trendall, eds., *Patterns of Change in Earth Evolution,* Springer (1984); T. Fenchel and B. J. Finlay, *Ecology and Evolution in Anoxic Worlds,* Oxford University Press (1995), pp. 63–66, 73–77; and K. Konhauser, *Introduction to Geomicrobiology,* Blackwell (2007). Lively accounts of the origin of oxygenic photosynthesis are given by O. Morton, *Eating the Sun: How Plants Power the Planet,* HarperCollins (2007), pp. 145–228 and N. Lane, *Life Ascending,* pp. 60–87.

The fundamental processes of respiration include aerobic respiration, whose basic formula is $C_6H_{12}O_6 + 6O_2 \rightarrow 6CO_2 + 6H_2O$. Burning sugar with fire yields the same reaction, as the chemists Priestley and Lavoisier discovered. Respiration, however, is a complex process that strips hydrogen atoms from sugar and transfers them by stages to where they combine with oxygen to make water. Some anaerobic bacteria use sulphate, SO_4^{2-}, not oxygen, O_2, to extract energy from sugar:

$$6H^+ + 3SO_4^{2-} + C_6H_{12}O_6 \rightarrow 6CO_2 + 3H_2S + 6H_2O.$$

These "bacterial sulphate reducers," which evolved at least 3.5 billion years ago, keep the earth from being coated with a skin of gypsum (calcium sulphate, $CaSO_4$), but 1.8 billion years ago, sulphides released by volcanoes, washed into the oceans, and subsequently oxidized, rendered subsurface ocean waters sulphidic and anoxic, as is true in the Black Sea today. Sulphate reducers outcompete methanogens where sulphates are available, for they, too, can use hydrogen to extract energy from sulphate (Fenchel & Finlay, *Ecology and Evolution in Anoxic Worlds,* p. 64): $4H_2 + 2H^+ + SO_4^{2-} \rightarrow H_2S + 4H_2O$. Notice that if oxygen were available, further energy could be obtained from the reaction

$$2H_2S + O_2 \rightarrow 2H_2O + 2S.$$

This reaction provides the energy for the life that teems around "black smokers." Other bacteria extract energy from methane with sulphate (Konhauser, 2007, p. 83):

$$CH_4 + SO_4^{2-} \rightarrow HCO_3^- + HS^- + H_2O.$$

Yet others use nitrates to extract energy from sugar, according to the reaction

$$24H^+ + 24NO_3^- + 5C_6H_{12}O_6 \rightarrow 12N_2 + 30CO_2 + 42H_2O.$$

Fermenters of carbohydrates break down glucose into pyruvate ($CH_3COCOOH$) and hydrogen, and then convert pyruvate into vinegar, carbon dioxide, and hydrogen:

$$C_6H_{12}O_6 \rightarrow 2CH_3COCOOH + 2H_2, 2CH_3COCOOH + 2H_2O \rightarrow 2CH_3COOH + 2CO_2 + 2H_2.$$

Light is used to "photosynthesize" energy-rich compounds in various ways: Green sulphur bacteria use bacteriophyll, a primitive analogue of chlorophyll, to trap the light needed to make sugar by combining hydrogen sulphide, H_2S, and carbon dioxide:

$$6CO_2 + 12H_2S \rightarrow C_6H_{12}O_6 + 12S + 6H_2O,$$

$$6CO_2 + 6H_2O + 3H_2S \rightarrow C_6H_{12}O_6 + 3SO_4^{2-} + 6H^+$$

(Konhauser, *Introduction to Geomicrobiology*, p. 51). They can use elemental sulphur to make sugar (p. 52):

$$4S + 6CO_2 + 10H_2O \rightarrow C_6H_{12}O_6 + 8H^+ + 4SO_4^{2-}$$

Iron bacteria use light trapped by bacteriophyll to combine ferrous iron (Fe^{2+}), carbon dioxide, and water to make sugar, ferric hydroxide, and protons (H^+) (Konhauser, p. 52):

$$6CO_2 + 24Fe^{2+} + 66H_2O \rightarrow C_6H_{12}O_6 + 24Fe(OH)_3 + 48H^+.$$

Other bacteria use trapped light to make sugar by oxidizing manganese:

$$12Mn^{2+} + 6CO_3^{2-} + 12H_2O \rightarrow 12MnO_2 + C_6H_{12}O_6 + 12H^+$$

Cyanobacteria use oxygen photosynthesis to make sugar from carbon dioxide and water, employing energy from photons trapped in their chlorophyll, by the reaction

$$6CO_2 + 12\ H_2O \rightarrow C_6H_{12}O_6 + 6O_2 + 6H_2O.$$

Atmospheric oxygen was half the modern level 800 million years ago: N. J. F. Blamey et al., "Paradigm shift in determining Neoproterozoic atmospheric oxygen," *Geology,* vol. 44, pp. 651–654 (2016). Deeper ocean water did not oxygenate until sponges began clearing the water of organic particles, mostly dead cyanobacteria, whose decay consumed O_2: T. M. Lenton et al., "Co-evolution of eukaryotes and ocean oxygenation in the Neoproterozoic era," *Nature Geoscience,* vol. 7, pp. 257–265 (2014) and N. J. Butterfield, "The Neoproterozoic," *Current Biology,* vol. 25, pp. R859–R863 (2015).

DIVERSIFICATION AND EVOLUTIONARY PROGRESS

The initial quotation is from J. B. S. Haldane and J. S. Huxley, *Animal Biology,* Oxford University Press (1927), p. 234. They present a decisive argument for evolutionary progress, in the sense of a tendency to technological improvement among dominant organisms, in their fig. 117 (pp. 324–325).

Making Complex, Higher Organisms: Turning Bacteria into Live-In Power Plants

In *Symbiosis in Cell Evolution,* W. H. Freeman (1993), L. Margulis showed that mitochondria and chloroplasts descend from independent organisms. The account given here of the taming of mitochondria follows N. Lane, *The Vital Question,* Norton (2015), pp. 157–191. T. M. Embley and T. A. Williams list characteristics of an archaean like the one that domesticated the ancestors of mitochondria in "Steps on the road to eukaryotes," *Nature,* vol. 521, pp. 169–170 (2015). Lane discusses the consequences of this event in *The Vital Question,* pp. 192–233. N. J. Butterfield describes how large active animals sped up evolution in "Animals and the invention of the Phanerozoic Earth system," *Trends in Ecology and Evolution,* vol. 26, pp. 81–87 (2011).

F. B. Essig shows how different groups of organisms obtained chloroplasts in *Plant Life: A Brief History,* Oxford University Press (2015), pp. 25–30.

Making Complex, Higher Organisms: Origins of Complex Many-Celled Organisms

How multicellularity evolved is discussed by Maynard Smith and Szathmáry in *The Origins of Life,* Oxford University Press (1999), pp. 109–124; R. K. Grosberg and R. Strathmann, "The evolution of multicellularity: A minor major transition?" *Annual Review of Ecology, Evolution and Systematics,* vol. 38, pp. 621–654 (2007); and A. H. Knoll, "The multiple origins of complex multicellularity," *Annual Review of Earth and Planetary Sciences,* vol. 39, pp. 217–239 (2011).

The Origin and Diversification of Mobile, Sentient Many-Celled Animals

This section is based on G. J. Vermeij, "Inequality and the directionality of history," *American Naturalist,* vol. 153, pp. 243–254 (1999); N. J. Butterfield, "Oxygen, animals, and aquatic bioturbation, an updated account," *Geobiology,* vol. 16, pp. 3–16 (2018); the early chapters of D. H. Erwin and J. W. Valentine's comprehensive analysis of Cambrian diversification, *The Cambrian Explosion,* Roberts & Co. (2013); C. W. Dunn et al., "The hidden biology of sponges and ctenophores," *Trends in Ecology and Evolution,* vol. 30, pp. 282–291 (2015); H. Philippe et al., "Acoelomorph flatworms are deuterostomes related to *Xenoturbella,*" *Nature,* vol. 470, pp. 255–258 (2011); O. Rota-Stabelli et al., "Molecular timetrees reveal a Cambrian colonization of land and a new scenario for ecdysozoan evolution," *Current Biology,* vol. 23, pp. 392–398 (2013); and M. F. Land, *The Eye: A Very Short Introduction,* Oxford University Press (2014), p. 4. Common ancestors of vertebrates, arthropods, and annelids had central nervous systems and embryonic brains, and some annelids have precursors of notochords programmed by homologous genes: see D. Arendt and K. Nübler-Jung, "Dorsal or ventral: Similarities in fate maps and gastrulation patterns in annelids, arthropods and chordates," *Mechanisms of Development,* vol. 61, pp. 7–21 (1997) and D. Arendt et al., "The evolution of nervous system centralization," *Philosophical Transactions of the Royal Society,* series B, vol. 363, pp. 1523–1528 (2008). M. W. Denny discusses spider webs in "The physical properties of spider's silk and their role in the design of orb-webs," *Journal of Experimental Biology,* vol. 65, pp. 483–506 (1976); S. A. Combes outlines surprising abilities of dragonflies in "Dragonflies predict and plan their hunts," *Nature,* vol. 517, pp. 279–280 (2015). Outstanding features of cephalopods are outlined in S. Conway Morris, *Life's Solutions,* Cambridge University Press (2003), pp. 151–154 and 214–217; P. Godfrey-Smith, *Other Minds: The Octopus, the Sea, and the Deep Origins of Consciousness,* Farrar, Straus and Giroux (2016); and M. Moynihan, *Communication and Noncommunication by Cephalopods,* Indiana University

Press (1985), illustrated by the author's superb art. R. M. Alexander describes tunicates (sea squirts) in *The Chordates,* 2nd ed., Cambridge University Press (1981), pp. 25–30.

Our account of the Ediacaran and Cambrian is based on D. H. Erwin and J. W. Valentine, *The Cambrian Explosion,* Roberts & Co. (2013). In contrast to Erwin and Valentine, however, Gregory Retallack, "Ediacaran life on land," *Nature,* vol. 493, pp. 89–92 (2013) believes that many Ediacaran organisms were terrestrial or intertidal lichens. A. Y. Zhuravlev et al. describe a possible bryozoan ancestor in "Ediacaran skeletal metazoan interpreted as a lophophorate," *Proceedings of the Royal Society,* series B, vol. 282, article #20151860 (2015); R. M. Alexander describes lophophorates like bryozoans in *The Invertebrates,* Cambridge University Press (1979), pp. 485–495. C. L. Smith et al. describe *Trichoplax* behavior in "Coordinated feeding behavior in *Trichoplax,* an animal without synapses," *PLoS One,* vol. 10 (issue 9), e0136098 (2015). N. J. Butterfield describes the modernization of Cambrian ecosystems in "Animals and the invention of the Phanerozoic Earth system," *Trends in Ecology and Evolution,* vol. 26, pp. 81–87 (2015). Stages in the development of animals' awareness of and ability to respond to their environment are described by X. Ma et al., "Complex brain and optic lobes in an early Cambrian arthropod," *Nature,* vol. 490, pp. 258–261 (2012) and J. R. Paterson et al., "Acute vision in the giant Cambrian predator *Anomalocaris* and the origin of compound eyes," *Nature,* vol. 480, pp. 237–240 (2011). D. Floreano et al. discuss what biologists can learn from "intelligent robots" in "Robotics and neuroscience," *Current Biology,* vol. 24, pp. R910–R920 (2014). D. A. T. Harper describes the great Ordovician diversification in "The Ordovician biodiversification: Setting an agenda for marine life," *Palaeogeography, Palaeoclimatology, Palaeoecology,* vol. 232, pp. 148–166 (2006).

Trade-Offs and Diversification

Robert MacArthur explained trade-offs in "Population consequences of natural selection," *American Naturalist,* vol. 95, pp. 195–199 (1961), and Ronald Fisher (without naming them) explained their role in speciation in *Genetical Theory of Natural Selection,* Oxford University Press (1930), pp. 126–128. D. Ross Robertson demonstrated a trade-off between defending rich resources versus surviving scarcity in "Interspecific competition controls abundance and habitat use of territorial Caribbean damselfishes," *Ecology,* vol. 77, pp. 885–899 (1996). J. L. Wulff outlines the trade-off sponges face between fast growth and effective defense in "Trade-offs in resistance to competitors and predators, and their effects on the diversity of tropical marine sponges," *Journal of Animal Ecology,* vol. 74, pp. 313–321 (2005).

TRADE-OFFS AND COMMUNITY ORGANIZATION

Near-Shore Marine Communities: An Organizing Trade-Off

In the nicely illustrated *Intertidal Wilderness,* University of California Press, Berkeley (2002), A. W. Rosenfeld and R. T. Paine describe the intertidal community at Tatoosh. R. T. Paine and R. L. Vadas discuss the impact of sea urchins in "The effects of grazing by sea urchins, *Strongylocentrotus* spp., on benthic algal populations," *Limnology and Oceanography,* vol. 14, pp. 710–719 (1969). R. T. Paine discusses the ecology of sea palms in a paper in a journal with a lovely cover picture of sea palms, "Disaster, catastrophe, and local persistence in the sea palm *Postelsia palmaeformis,*" *Science* vol. 205, pp. 685–687 (1979). E. G. Leigh et al. describe the effect of wave action and intertidal productivity in "Wave action and intertidal productivity," *Proceedings of the National Academy of Sciences, USA,* vol. 84, pp. 1314–1318 (1987).

Trade-Offs and Marine Communities on Panama's Two Coasts

J. B. C. Jackson and L. D'Croz describe the contrast between Panama's two coasts in "The oceans divided," in A. G. Coates, ed., *Central America: A Natural and Cultural History,* Yale University Press (1997), pp. 38–71. L. G. Gillis et al. describe the interdependence between coral reefs and the seagrass beds and mangroves they shelter from ocean waves in "Potential for landscape-scale positive interactions among tropical marine ecosystems," *Marine Ecology Progress Series,* vol. 503, pp. 289–303 (2014). C. Wilkinson and A. Cheshire compare near- and offshore sponges in "Comparisons of sponge populations across the Barrier Reefs of Australia and Belize: Evidence for higher productivity in the Caribbean," *Marine Ecology Progress Series,* vol. 67, pp. 283–294 (1990). C. Birkeland describes the contrast between fast growth in upwelling ecosystems and effective defense in coral reef communities in Panama, in "The importance of rate of biomass accumulation in early successional stages of benthic communities to the survival of coral recruits," in *Proceedings, Third International Coral Reef Symposium,* Rosenstiel School of Marine and Atmospheric Science, Miami, Florida (1977), pp. 15–21, "Geographic comparisons of coral reef community processes," in *Proceedings of the 6th International Coral Reef Symposium, Australia,* vol. 1 (1988), pp. 211–220, and "Symbiosis, fisheries and economic development on coral reefs," *Trends in Ecology and Evolution,* vol. 12, pp. 364–367 (1997).

How to Detect an Organizing Trade-Off from the Fossil Record

Aaron O'Dea and Beth Okamura show how to deduce seasonal cooling from variation in zooid size of bryozoans in "Intracolony variation in zooid size in cheilostome bryozoans as a new technique for investigating paleoseasonality," *Palaeogeography, Palaeoclimatology, Palaeoecology,* vol. 162, pp. 319–332 (2000). A. O'Dea et al. use this method to show that Caribbean upwellings declined after the Isthmus of Panama formed in "Environmental change preceded Caribbean extinction by 2 million years," *Proceedings of the National Academy of Sciences, USA,* vol. 104, pp. 5501–5506 (2007).

THE EVOLUTION OF LAND PLANTS

Trade-Offs Between Living in Water and Living on Land

G. J. Retallack et al. describe a soil microbe community that lived on a coastal plain three billion years ago in "Archaean coastal-plain paleosols and life on land," *Gondwana Research,* vol. 40, pp. 1–20 (2016).

K. J. Willis and J. C. McElwain discuss the adaptations needed for land plants to evolve in *The Evolution of Plants,* 2nd edition, Oxford University Press (2014), pp. 61–71.

How Plants Adjusted to Living on Land

The ancestors of land plants, and how early plants adjusted to land, are discussed by P. Kenrick and P. R. Crane, "The origin and early evolution of plants on land," *Nature,* vol. 389, pp. 33–39 (1997); K. J. Willis and J. C. McElwain, *The Evolution of Plants,* pp. 72–91; P. Steemans et al., "Origin and radiation of the earliest vascular land plants," *Science,* vol. 324, p. 353 (2009); R. Ligrone et al., "Major transitions in the evolution of early land plants: A bryological perspective," *Annals of Botany,* vol. 109, pp. 851–871 (2012); J. Pitterman, "The evolution of water transport in plants: An integrated approach," *Geobiology,* vol. 8, pp. 112–139 (2010); and T. J. Brodribb and S. A. M. MacAdam, "Evolution of the stomatal regulation of plant water content," *Plant Physiology,* vol. 174, pp. 639–649 (2017). The quotation from E. J. H. Corner about small sperm versus larger eggs is from *The Life of Plants,* World Press (1964), p. 86. M. I. Bidartondo et al. show how symbiosis with mycorrhiza-like fungi enabled plants to colonize the land in "The dawn of symbiosis between plants and fungi," *Biology Letters,* vol. 7, pp. 574–577 (2011). The role of early plants in triggering late Ordovician glaciations is discussed by T. M. Lenton et al., "First plants cooled the Ordovician," *Nature Geoscience,* vol. 5, pp. 86–89 (2012) and G. J. Retallack, "Late Ordovician glaciation initiated by early land plant evolution and punctuated by greenhouse mass extinctions," *Journal of Geology*, vol. 123, pp. 509–538 (2015).

D. Rota-Stabelli et al. show when different animal groups first colonized the land in "Molecular timetrees reveal a Cambrian colonization of land and a new scenario for ecdysozoan evolution," *Current Biology*, vol. 23, pp. 392–398 (2013). Conrad Labandeira describes the early diversification of arthropod herbivores in "Silurian to Triassic plant and hexapod clades and their associations: New data, a review, and interpretations," *Arthropod Systematics and Phylogeny*, vol. 64, pp. 53–94 (2006), "The four phases of plant-arthropod association in deep time," *Geologica Acta*, vol. 4, pp. 409–438 (2006), and "The origin of herbivory on land: Initial patterns of plant tissue consumption by arthropods," *Insect Science*, vol. 14, pp. 259–275 (2007). C. K. Boyce discusses *Cooksonia* in "How green was *Cooksonia* . . . ," *Paleobiology*, vol. 34, pp. 179–194 (2008).

The Evolution of Trees

The evolution of trees and forest is described in G. J. Retallack, *Soils of the Past*, Blackwell (2001), pp. 280–299 and K. J. Willis and J. C. McElwain, *The Evolution of Plants*, 2nd edition, Oxford University Press (2014), pp. 113–178. Melvin Tyree and Martin Zimmermann show how transpiration works in *Xylem Structure and the Ascent of Sap*, Springer (2002). The quotation about transpiration is from Corner, *The Life of Plants* (1964), p. 107. Willis and McElwain discuss how seeds evolved in *The Evolution of Plants*, pp. 109–116. D. Floudas et al. date the origin of lignin decomposition in "The Paleozoic origin of enzymatic lignin decomposition reconstructed from 31 fungal genomes," *Science*, vol. 336, pp. 1715–1719 (2012). G. Retallack shows how to infer the atmosphere's carbon dioxide content from stomatal index in "A three-hundred-million year record of atmospheric carbon dioxide content from fossil plant cuticles," *Nature*, vol. 411, pp. 287–290 (2001). Peter Crane discusses ginkgos in *Ginkgo*, Yale University Press (2013). The quotation about the gloominess of conifer forests is from E. J. H. Corner, *The Life of Plants*, World Press (1964), p. 117.

The Evolution of Flowering Plants

The trade-off between fast growth and anti-herbivore defense is described by P. D. Coley et al., "Resource availability and plant herbivore defense," *Science*, vol. 230, pp. 895–899 (1985). E. G. Leigh, Jr., discusses the role of animal pollinators in the spread and expansion of flowering plants on p. 2515 of "The evolution of mutualism," *Journal of Evolutionary Biology*, vol. 23, pp. 2507–2528 (2010).

Conrad Labandeira et al. describe early examples of insect pollination in

"Pollination drops, pollen, and insect pollination of Mesozoic gymnosperms," *Taxon,* vol. 56, pp. 663–695 (2007), mentioning the pollination mechanism in cycads and ginkgos on p. 666. Labandeira revisits the topic in "The pollination of mid-Mesozoic seed plants and the early history of long-proboscid insects," *Annals of the Missouri Botanical Garden,* vol. 97, pp. 469–513 (2010). C. Labandeira et al. describe mid-Jurassic butterfly-like pollinators in the online article "The evolutionary convergence of mid-Mesozoic lacewings and Cenozoic butterflies," *Proceedings of the Royal Society,* series B, vol. 283, article #20152893 (2016). Cycads are slow-growing and poisonous: A. Prado et al., "Leaf traits and herbivory levels in a tropical gymnosperm, *Zamia stevensonii* (Zamiaceae)," *American Journal of Botany,* vol. 101, pp. 437–447 (2014).

Nathan Jud and Leo Hickey suggest that angiosperms began as herbs in "*Potomacapnos apeleutheron* gen and sp. nov., a new early Cretaceous angiosperm from the Potomac Group and its implications for the evolution of eudicot leaf architecture," *American Journal of Botany,* vol. 100, pp. 2437–2449 (2013): also see N. Jud, "Fossil evidence for a herbaceous diversification of early eudicot angiosperms during the Early Cretaceous," *Proceedings of the Royal Society,* series B, vol. 282, article #20151045 (2015). The increase in leaf venation 100 million years ago and its consequences are described in C. K. Boyce et al., "Angiosperm leaf vein evolution was physiologically and environmentally transformative," *Proceedings of the Royal Society,* series B, vol. 276, pp. 1771–1776 (2009); C. K. Boyce and J.-E. Lee, "An exceptional role for flowering plant physiology in the expansion of tropical rainforests and biodiversity," *Proceedings of the Royal Society,* series B, vol. 277, pp. 3437–3443 (2010); and T. S. Field et al., "Fossil evidence for Cretaceous escalation in angiosperm leaf vein evolution," *Proceedings of the National Academy of Sciences, USA,* vol. 108, pp. 8363–8366 (2011). Vein density is a far from perfect measure of photosynthetic capacity: F. E. Rockwell and N. M. Holbrook, "Leaf hydraulic architecture and stomatal conductance: A functional perspective," *Plant Physiology,* vol. 174, pp. 1996–2007 (2017). Sadly, this paper is quite difficult.

The combination of animal pollination and higher photosynthesis in flowering plants allowed terrestrial to surpass marine diversity in an unprecedented manner: G. J. Vermeij and R. K. Grosberg, "The great divergence: When did diversity on land exceed that in the sea?" *Integrative and Comparative Biology,* vol. 50, pp. 675–682 (2010).

How trees manage water is discussed by J. S. Sperry et al., "Safety and efficiency conflicts in hydraulic architecture: Scaling from tissues to trees," *Plant, Cell and Environment,* vol. 31, pp. 632–645 (2008); F. C. Meinzer et al., "Xylem hydraulic safety margins in woody plants: Coordination of stomatal control of xylem tension with hydraulic capacitance," *Functional Ecology,* vol. 23, pp. 922–930 (2009);

J. Pitterman, "The evolution of water transport in plants: An integrated approach," *Geobiology,* vol. 8, pp. 112–139 (2010); T. J. Brodribb et al., "Conifer species adapt to low-rainfall climates by following one of two divergent pathways," *Proceedings of the National Academy of Sciences, USA,* vol. 111, pp. 14489–14493 (2014); and T. J. Brodribb and S. A. M. MacAdam, "Evolution of the stomatal regulation of plant water content," *Plant Physiology,* vol. 174, pp. 639–649 (2017).

E. J. H. Corner describes the role of cooperation between plants and animals in the evolution of flowering forest in *The Life of Plants,* p. 185. The contrast in how gymnosperm and angiosperms transport water from roots to and throughout leaves are outlined in J. Pittermann et al., "Torus-margo pits help conifers compete with angiosperms," *Science,* vol. 310, p. 1924 (2005), and T. J. Brodribb & T. S. Field, "Leaf hydraulic evolution led a surge in leaf photosynthetic capacity during early angiosperm diversification," *Ecology Letters,* vol. 13, pp. 175–183 (2010).

Chapter 6. Integrating Diversity into Community: Interdependence and Mutualism

HUMAN ECONOMIES: HOW COMPETITION FAVORS DIVERSIFICATION AND INTERDEPENDENCE

The parallels between human economies and natural ecosystems are explored by G. J. Vermeij, *Nature: An Economic History,* Princeton University Press (2004); and G. J. Vermeij & E. G. Leigh, Jr., "Natural and human economies compared," *Ecosphere,* volume 2(4), Article 39 (2011). The quotation from Adam Smith is from *An Enquiry into the Nature and Causes of the Wealth of Nations,* Strahan and Cadell (1776), Book I, chapter 1, ¶ 11. The pin-making example is from Book I, chapter 1, ¶ 3. The importance for an economy's health of the fairness of competition is outlined by Adam Smith in *The Theory of Moral Sentiments,* A. Millar (1759), Part II, Section II, chapter 3, ¶ 3.

PRODUCTIVITY AND ITS HINDRANCES

How we know ecosystems are organized for high productivity and diversity is explained by E. G. Leigh, Jr., and G. J. Vermeij in "Does natural selection organize ecosystems for the maintenance of high productivity and diversity?" *Philosophical Transactions of the Royal Society of London,* series B, vol. 357, pp. 709–718 (2002). James Estes provides further evidence that novel human disturbance can radically reduce the productivity and diversity of natural ecosystems (without comment or conclusion!) in his fine book, *Serendipity: An Ecologist's Quest to Understand Na-*

ture, University of California Press (2016). Productivity of and light interception by tropical forest is discussed in *Tropical Forest Ecology,* pp. 120–148. Garrett Hardin described "tragedies of the commons" in "The tragedy of the commons," *Science,* vol. 162, pp. 1243–1248 (1968). Leigh outlines how such tragedies limit forest production on p. 124 of "Tropical forest ecology: Sterile or virgin for theoreticians?" in W. P. Carson and S. A. Schnitzer, eds., *Tropical Forest Community Ecology,* Wiley-Blackwell, pp. 121–142 (2008). S. C. Sillett and T. van Pelt estimate the dry weight of redwood wood and leaf in a hectare of lowland redwood forest in "Trunk reiteration promotes epiphytes and water storage in an old-growth redwood forest canopy," *Ecological Monographs,* vol. 77, pp. 335–359 (2007). Other disadvantages of competing for light are outlined in G. W. Koch et al., "The limits to tree height," *Nature,* vol. 428, pp. 851–854 (2004); J.-C. Domec et al., "Maximum height in a conifer is associated with conflicting requirements for xylem design," *Proceedings of the National Academy of Sciences, USA,* vol. 105, pp. 12069–12074 (2008), and Leigh, "Tropical forest ecology: Sterile or virgin for theoreticians?" pp. 123–125.

FAVORABLE CLIMATES INTENSIFY COMPETITION, DIVERSITY, AND INTERDEPENDENCE

Theodosius Dobzhansky outlined the relation between favorable tropical climate and intense competition among tropical plants and animals, and the consequences of this competition for tropical diversity, in "Evolution in the tropics," *American Scientist,* vol. 38, pp. 209–221 (1950).

Table 1, data for Harvard Forest from R. H. Waring et al., "Scaling gross ecosystem production at Harvard Forest . . . ," *Plant, Cell and Environment,* vol. 18, pp. 1201–1213 (1995), table 4; data for the other two forests from Table 8.1, p. 125 of Leigh, "Tropical forest ecology . . ." in W. P. Carson and S. A. Schnitzer, eds., *Tropical Forest Community Ecology,* Wiley-Blackwell (2008). I assume a ton of photosynthesized carbon is 2.5 tons of photosynthesized starch or sugar.

How intensely tropical plants compete for light is illustrated by Christian Ziegler and Egbert Leigh in *A Magic Web,* Oxford University Press (2002), pp. 100–121 and documented in *Tropical Forest Ecology,* p. 126. R. S. Beeman et al. show how hard plants can work to attract pollinators in "Pollination of *Rafflesia* (Raffleciaceae)," *American Journal of Botany,* vol. 75, pp. 1148–1162 (1988). E. G. Leigh et al. show how competition breeds interdependence in "What do human economies, large islands and forest fragments reveal about the factors limiting ecosystem evolution?" *Journal of Evolutionary Biology,* vol. 22, pp. 1–12 (2009). Species counts for insect-eating birds in South Carolina and the Manú are calculated from J. Terborgh, *Diversity and the Rain Forest,* Scientific American Library

(1992), p. 64; the count for French Guiana is calculated from J. M. Thiollay, "Structure, density and rarity in an Amazonian forest bird community," *Journal of Tropical Ecology,* vol. 10, pp. 449–481 (1994). Tropical predators' impact on their prey is outlined in *A Magic Web,* pp. 166–193.

Why pest pressure is higher in the tropics, and why intense pest pressure allows so many kinds of tropical trees to coexist, especially in everwet climates, are outlined in D. H. Janzen, "Herbivores and the number of tree species in tropical forests," *American Naturalist,* vol. 104, pp. 501–528 (1970); Leigh, *Tropical Forest Ecology,* pp. 190–194; and C. Ziegler and E. G. Leigh, *A Magic Web,* Oxford University Press (2002), pp. 50–57. The consumption rates of young leaves on Barro Colorado Island relative to temperate-zone counterparts are calculated from Table 1 of P. D. Coley and J. A. Barone, "Herbivory and plant defenses in tropical forests," *Annual Review of Ecology and Systematics,* vol. 27, pp. 305–335 (1996), assuming that leaves at all latitudes take equal time to mature. T. M. Aide and E. C Londoño discuss *Gustavia*-eating caterpillars in "The effects of rapid leaf expansion on the growth and survivorship of a lepidopteran herbivore," *Oikos,* vol. 55, pp. 66–70 (1988). S. Koptur shows that ants defend young *Inga* leaves in "Experimental evidence for defense of *Inga* (Mimosoideae) saplings by ants," *Ecology,* vol. 65, pp. 1787–1793 (1984). D. H. Janzen discusses ants that live by defending acacia trees in "Coevolution of mutualism between ants and acacias in Central America," *Evolution,* vol. 20, pp. 249–275 (1966); and ants that defend *Cecropia* trees in "Allelopathy by myrmecophytes: The ant *Azteca* as an allelopathic agent of *Cecropia,*" *Ecology,* vol. 50, pp. 147–153 (1969). Phyllis Coley and Thomas Kursar discuss different ways plants defend young leaves in "Anti-herbivore defenses of young tropical leaves: Physiological constraints and ecological trade-offs," in S. S. Mulkey et al., eds., *Tropical Forest Plant Ecophysiology,* Chapman and Hall (1996), pp. 305–336. Phyllis Coley et al. show how this work singles out young leaves most likely to harbor chemicals of medical interest in "Use of ecological criteria in designing plant collection strategies for drug discovery," *Frontiers in Ecology and the Environment,* vol. 1, pp. 421–428 (2003). Pests devastate single-species stands of plants: H. N. Ridley, *The Dispersal of Plants Throughout the World,* Reeve (1930), p. xvi.

Liza Comita et al. show that seedlings do less well where there are more adults of their species nearby in "Asymmetric density dependence shapes species abundances in a tropical tree community," *Science,* vol. 329, pp. 330–332 (2010). K. S. Murali and R. Sukumar show that plants flush leaves before the rains to avoid insect attack in "Leaf flushing phenology and herbivory in a tropical dry deciduous forest, southern India," *Oecologia,* vol. 94, pp. 114–119 (1993). T. A. Kursar et al. show that coexisting, closely related tree species differ only in anti-pest defenses in "The evolution of antiherbivore defenses and their contribution to species co-existence in the tropical tree genus *Inga*," *Proceedings of the National Academy of Sciences, USA,* vol. 106, pp. 18073–18078 (2009). D. H. Janzen describes caterpillar diets and insect outbreaks in "Ecological characterization of a Costa Rican dry forest caterpillar fauna," *Biotropica,* vol. 20, pp. 120–135 (1988). R. Bagchi et al., however, showed that spraying insecticide on the forest floor reduces seedling diversity only slightly in "Pathogens and insect herbivores drive rainforest plant diversity and composition," *Nature,* vol. 506, pp. 85–88 (2014).

How effectively birds and bats protect tropical forests from insect pests is outlined in Leigh, *Tropical Forest Ecology,* pp. 164–168 and M. B. Kalka et al., "Bats limit arthropods and herbivory in a tropical forest," *Science,* vol. 320, p. 71 (2008). The daily energy expenditure of tropical birds is only 83 percent of that predicted by the formula Leigh used to calculate the energy expenditure of a bird of given weight: see Fig. 1 of K. J. Anderson and W. Jetz, "The broad-scale ecology of energy expenditure of endotherms," *Ecology Letters,* vol. 8, pp. 310–318 (2005). Marco Visser et al. outline how predators can abolish the density dependence of insect impacts in "Tri-trophic interactions affect density dependence of seed fate in a tropical forest palm," *Ecology Letters,* vol. 14, pp. 1093–1100 (2011).

In Table 2, data for central Panama (Barro Colorado Island) are from Appendix 8.1 of Leigh, *Tropical Forest Ecology;* data for forest in Ecuador and SE Peru are from p. 2104 of N. C. A. Pitman et al., "Dominance and distribution of tree species in upper Amazonian terra firme forests," *Ecology,* vol. 82, pp. 2101–2117 (2001); data for Costa Rican forest are from R. J. Burnham, "Stand characteristics and leaf litter composition of a dry forest hectare in Santa Rosa National Park, Costa Rica," *Biotropica,* vol. 29, pp. 384–395 (1997); Professor Hao Zhangqing provided data on the Manchurian forest.

Scott Mangan et al. show how soil-borne diseases affect the abundances of tropical trees in "Negative plant-soil feedback predicts tree-species relative abundance in a tropical forest," *Nature,* vol. 466, pp. 752–755 (2010). G. J. Vermeij shows how successive ecological dominants intensify competition over time in "Inequality and the directionality of history," *American Naturalist,* vol. 153, pp. 243–253 (1999).

COOPERATING THE BETTER TO COMPETE

E. O. Wilson shows how social insects shape ecosystems in *The Insect Societies,* Harvard University Press (1971). David Roubik describes the neotropical invasion of Africanized honeybees on p. 335 and pp. 357–366 of *Ecology and Natural History of Tropical Bees,* Cambridge University Press (1989).

Mutualisms Between Trees and Their Pollinators and Seed Dispersers

Leigh explains how pollinators promote tree diversity and anti-herbivore defense in *Tropical Forest Ecology,* pp. 190–194, 211–212. Christian Ziegler and Egbert Leigh describe some mutualisms between members of different species, including pollination and seed dispersal, in *A Magic Web,* Oxford University Press (2002), pp. 62–81; Leigh discusses the subject more comprehensively in "The evolution of mutualism," *Journal of Evolutionary Biology,* vol. 23, pp. 2507–2528 (2010). The mutualism between fig trees and their pollinating wasps is outlined in *A Magic Web,* pp. 77–81. E. O. Martinson et al. show that fig wasps sting ovules to make them suitable homes for their larvae in "Relative investment in egg load and poison sac in fig wasps: Implications for physiological mechanisms underlying seed and wasp production in figs," *Acta Oecologica,* vol. 57, pp. 58–66 (2014). E. J. H. Corner's remark is from p. 24 of "*Ficus* in the Solomon Islands, and its bearing on the post-Jurassic history of Melanesia," *Philosophical Transactions of the Royal Society of London,* series B, vol. 253, pp. 23–159 (1967). C. Korine et al. show how many bat species depend on fig fruit in "Fruit characteristics and factors affecting fruit removal in a Panamanian community of strangler figs," *Oecologia,* vol. 123, pp. 560–568 (2000). Scott Wing and Lisa Boucher describe the spread of flowering plants, and its consequences, in "Ecological aspects of the Cretaceous flowering plant radiation," *Annual Review of Earth and Planetary Sciences,* vol. 26, pp. 379–421 (1998). Alison Jolly discusses intelligence and why it evolved in a beautifully written paper, "Lemur social behavior and primate intelligence," *Science,* vol. 153, pp. 501–506 (1966) and on pp. 366–367 of *The Evolution of Primate Behavior,* 2nd ed, Macmillan (1985).

Mutualism and the Spread of Coral Reefs

R. Wood outlines the history of coral reefs in "Nutrients, predation and the history of reef-building," *Palaios,* vol. 8, pp. 526–543 (1993) and in "The ecological evolution of reefs," *Annual Review of Ecology and Systematics,* vol. 29, pp. 179–206 (2008). The role of photosynthetic algae in reef evolution is discussed in J. A. Talent, "Organic reef-building: Episodes of extinction and symbiosis?" *Senckenbergia Lethaia,* vol. 69, pp. 315–368 (1988); R. Cowen, "The role of algal symbiosis in reefs through time," *Palaios,* vol. 3, pp. 221–227 (1988); G. D. Stanley, "The history of early Mesozoic reef communities: A three-step process," *Palaios,* vol. 3, pp. 170–183 (1988) and "Photosymbionts and the evolution of modern coral reefs," *Science,* vol. 312, pp. 857–858 (2006); and J. H. Lipps and G. D. Stanley, Jr., "Photosymbiosis in past and present reefs," D. K. Hubbard et al., eds., *Coral Reefs at the Crossroads,* Springer (2016), pp. 47–68. Mutualisms between reef sponges and cyanobacteria are described in C. R. Wilkinson, "Interocean differences in size and nutrition of coral reef sponge populations," *Science,* vol. 236, pp. 1654–1657 (1987) and M. Oren et al., "Transmission, plasticity and the molecular identification of cyanobacterial symbionts in the Red Sea sponge *Diacarnus erythraenus,*" *Marine Biology,* vol. 148, pp. 35–41 (2005). Mutualisms between coral reef clams and brachiopods, and photosynthetic algae are described in R. Cowen, "Analogies between the recent bivalve *Tridacna* and the fossil brachiopods Lyttoniacea and Richthofeniacea," *Palaeogeography, Palaeoclimatology, Palaeoecology,* vol. 8, pp. 329–344 (1971) and A. Seilacher, "Patterns of Macroevolution through the Phanerozoic," *Palaeontology,* vol. 56, pp. 1273–1283 (2013). The extinctions ending the Permian and Triassic are described in P. B. Wignall, *The Worst of Times: How Life on Earth Survived Eighty Million Years of Extinctions,* Princeton University Press (2015); G. J. Retallack and A. H. Jahren, "Methane release from igneous intrusion of coal during late Permian extinction events," *Journal of Geology,* vol. 116, pp. 1–20 (2008); and Y. Sun et al., "Lethally hot temperatures during the early Triassic greenhouse," *Science,* vol. 338, pp. 366–370 (2012). Rudists appear to have been favored over corals by an uncommonly warm spell in the mid-Cretaceous: see p. 64 of N. M. Fraser et al., "Dissecting '*Lithiotis*' bivalves: Implications for the early Jurassic reef eclipse," *Palaios,* vol. 19, pp. 51–67 (2004). P. Schulte et al., "The Chicxulub asteroid impact and mass extinction at the Cretaceous-Paleogene boundary," *Science,* vol. 327, pp. 1214–1218 (2010) and P. R. Renne et al., "State shift in Deccan volcanism at the Cretaceous-Paleogene boundary possibly induced by impact," *Science,* vol. 350, pp. 76–78 (2015) discuss causes of the extinction ending the Cretaceous. Wignall shows that, despite great extrusions of lava then, no mass extinction ended the Paleocene in *The Worst of Times,* pp. 161–165. P. Wilf and C. Labandeira show that tree diversity increased then in "Responses of plant-insect interactions to Paleocene-Eocene warming," *Science,* vol. 284, pp. 2153–2156 (1999).

Turning Herbivores into Mutualists: The Coevolution of Grassland and Grazers

P. D. Coley and J. Barone compare the average proportion of a leaf's total damage inflicted in the first month of its life for tropical and temperate-zone settings in "Herbivory and plant defenses in tropical forests," *Annual Review of Ecology and Systematics,* vol. 27, pp. 305–335 (1996). Dry forest in south India supports 15 tons of mammals per square kilometer, compared with 1.3 in mildly seasonal forest at Manú, Peru (Leigh, *Tropical Forest Ecology,* p. 164, table 7.11). S. J. McNaughton shows how grassland anti-herbivore strategies contrast with forest, in "Ecology of a grazing ecosystem: The Serengeti," *Ecological Monographs,* vol. 55, pp. 259–294 (1985). Gregory Retallack discusses the evolution of grassland, and how, on large continents, grassland eventually replaced woodlands with annual rainfall less than 700 millimeters, in *Soils of the Past,* 2nd ed., Blackwell, pp. 300–316 (2001), and in pp. 283–285 of "Cenozoic paleoclimate on land in North America," *Journal of Geology,* vol. 115, pp. 271–294 (2007). Darwin shows how cattle suppress the growth of fir seedlings on p. 72 of *On the Origin of Species,* 1st ed., John Murray (1859). Using evidence from Retallack and McNaughton, Leigh shows how evolving grassland turned herbivores into mutualists in "The evolution of mutualism," *Journal of Evolutionary Biology,* vol. 23, pp. 2507–2528 (2010). Leigh et al. remark that grassland never evolved on islands smaller than New Zealand on pp. 135–136 of "The biogeography of large islands, or, how does the size of the ecological theater affect the evolutionary play?" *Revue d'Écologie,* vol. 62, pp. 105–168 (2007). In Australia, "living fossil" mallee woodland occupies settings with annual rainfall of 500–600 millimeters, as shown by G. J. Retallack, "Mallee model for mammal communities of the early Cenozoic and Mesozoic," *Palaeogeography, Palaeoclimatology, Palaeoecology,* vol. 342–343, pp. 111–129 (2012). S. J. McNaughton summarizes the ecology of east Africa's Serengeti grassland in "Ecology of a grazing ecosystem: The Serengeti," *Ecological Monographs,* vol. 55, pp. 259–294 (1985).

HOW FRAGMENTATION DIMINISHES COMPETITION, DIVERSITY, AND INTERDEPENDENCE

How the size of islands in reservoirs affects their ecosystems is discussed in E. G. Leigh, Jr., "Barro Colorado," pp. 88–91 in R. G. Gillespie and D. A. Clague, eds. *The Encyclopedia of Islands,* University of California Press (2009) and in E. G. Leigh et al., "What do human economies, large islands and forest fragments reveal about the factors limiting ecosystem evolution?" *Journal of Evolutionary Biology,* vol. 22, pp. 1–12 (2009). W. D. Robinson assesses loss of bird species from Barro Colorado since its isolation in "Long-term changes in the avifauna of Barro Colorado Island, Panama, a tropical forest isolate," *Conservation Biology,* vol. 13, pp. 85–97 (1999).

WHAT FACTORS CONSTRAIN ECOSYSTEM EVOLUTION?

How an island's size limits its ecosystem's properties is outlined by G. J. Vermeij in *The Economy of Nature,* Princeton University Press (2004), pp. 151–156, and by E. G. Leigh et al., "The biogeography of large islands, or, how does the size of the ecological theater affect the evolutionary play?" *Révue d'Écologie* (*La Terre et la Vie*), vol. 62, pp. 105–168 (2007), and "What do human economies, large islands and forest fragments reveal about the factors limiting ecosystem evolution?" *Journal of Evolutionary Biology,* vol. 22, pp. 1–12 (2009).

What Limits Diversity?

C. Mourer-Chaviré et al. list the thirty-three prehistoric species of land birds on Réunion in "The avifauna of Réunion Island . . . at the time of arrival of the first Europeans," *Smithsonian Contributions to Paleontology,* vol. 89, pp. 1–38 (1999).

What Limits the Intensity of Competition and the Pace of Life?

G. P. Burness et al. show how weight of the largest herbivore, and the largest carnivore, varies on land masses of different size in "Dinosaurs, dragons and dwarfs: The evolution of maximum body size," *Proceedings of the National Academy of Sciences, USA,* vol. 98, pp. 14518–14523 (2001). S. Carlquist discusses how isolation affects Hawaii's biota in *Hawaii: A Natural History,* Pacific Tropical Botanic Garden (1980), and how isolation affects other island biotas in *Island Life,* Natural History Press (1963). He discusses prickles on Hawaiian Lobelioideae, not shared by other Hawaiian plants, as a possible anti-herbivorous bird defense on p. 10 of his introduction to W. L. Wagner and V. A. Funk, eds. *Hawaiian Biogeography,* Smithsonian Institution Press, pp. 1–13 (1995).

Interdependence Is Weaker on Small, Isolated Natural Islands

Claire Micheneau et al. showed that orchids on the small island of Réunion employ far cheaper, less effective pollinators than did their ancestors in Madagascar in "Orchid pollination from Darwin to the present day," *Botanical Journal of the Linnean Society,* vol. 161, pp. 1–19 (2009).

How Isolation Weakens Competition and Makes It Easier for Exotics to Invade

Darwin explains why introduced species so often invade and spread widely in isolated islands or small continents in *On the Origin of Species,* John Murray (1859), pp. 106–107, 389–390. E. G. Leigh, Jr., et al. discuss the relationship of isolation, weak competition, and vulnerability to invaders in "The historical biography of the isthmus of Panama," *Biological Reviews,* vol. 89, pp. 148–172 (2014).

Chapter 7. Heredity, Natural Selection, and Evolution

How selection for helping relatives drove evolution of insect societies is described by W. D. Hamilton, "The genetical evolution of social behavior," parts I and II, *Journal of Theoretical Biology,* vol. 7, pp. 1–52 (1964), and T. D. Seeley, *The Wisdom of the Hive,* Harvard University Press (1995), pp. 1–16.

HOW MENDEL DEDUCED AND PROVED THE LAWS OF GENETIC INHERITANCE

The discussion of the principles of genetics is based on a translation of Gregor Mendel's (1866) paper, "Experiments in plant hybridization," reprinted in J. A. Peters, ed., *Classic Papers in Genetics,* Prentice-Hall (1959), pp. 1–20; and R. A. Fisher, *The Genetical Theory of Natural Selection,* Oxford University Press (1930), pp. 7–9. A different presentation is given by Brian Charlesworth and Deborah Charlesworth, *Evolution: A Very Short Introduction,* Oxford University Press (2003), pp. 24–34, who discuss genes and their arrangement on chromosomes on pp. 27–38 of the same book.

HOW ORGANISMS READ THEIR GENES' INSTRUCTIONS

James Watson et al. discuss genes, how they act, how their manufacture of enzymes is regulated and coordinated, and their arrangement on chromosomes in *The Molecular Biology of the Gene,* 4th ed., vol. 1, Benjamin/Cummings (1987). Jacques Monod gives a clearer but less up-to-date summary in *Chance and Necessity,* Vintage Books (1971), pp. 45–117. The role of the enzyme phosphoglucose isomerase in the life of sulphur butterflies, *Colias,* is explained in Ward Watt et al., "Females' choice of 'good genotypes' as mates is provided by an insect mating system," *Science,* vol. 233, pp. 1187–1190 (1986) and Christopher Wheat et al., "From DNA to fitness differences: Sequences and structures of adaptive variants of *Colias* phosphoglucose isomerase (PGI)," *Molecular Biology and Evolution,* vol. 23, pp. 499–512 (2006). M. W. Kirschner and J. C. Gerhart show how an animal's cells and tissues can differ when all have the same genes on pp. 209–215 of *The Plausibility of Life,* Yale University Press (2005); S. B. Carroll discusses why switch genes only work in certain places in *Endless Forms Most Beautiful: The New Science of Evo-Devo,* Norton (2005), see especially pp. 65–71, 92–93, and 116. The quotation is translated by Egbert Leigh from Georges Bernanos, "Nos amis les saints," Georges Bernanos, *La liberté pour quoi faire,* Gallimard (1953). Brian and Deborah Charlesworth outline how to infer the "family tree" of a particular group of plants or animals in *Evolution: A Very Short Introduction,* pp. 38–39, 101–109. The mammal phylogeny in fig. 128 is from M. dos Reis et al., "Phylogenetic datasets provide both precision and accuracy in estimating the timescale of placental mammal phylogeny," *Proceedings of the Royal Society,* series B, vol. 279, pp. 3491–3500 (2012).

EVOLUTION, NATURAL SELECTION, AND THE GENOME'S COMMON GOOD

Mark Ridley explains the importance of fair meiosis and the ways in which cheater genes can bias it in *Mendel's Demon [Gene Justice and the Diversity of Life],* Weidenfeld and Nicolson (2000).

Chapter 8. Organizing Genes for Adaptive Evolution

Charles Darwin remarks on p. 3 of *On the Origin of Species,* John Murray (1859) that he thought it inadequate to demonstrate the fact of evolution if he could not explain the origin of adaptation.

NATURAL SELECTION: SOME EXAMPLES

The specific mutations enabling a strain of bacteria to evolve resistance to a new drug are documented in D. M. Weinreich et al., "Darwinian evolution can follow only very few mutational paths to fitter proteins," *Science,* vol. 312, pp. 111–114 (2006). How selection drove the advance and retreat of the peppered moth's dark form is described on pp. 236–254 of Michael Majerus, *Moths,* HarperCollins (2002) and L. M. Cook et al., "Selective bird predation on the peppered moth: The last experiment of Michael Majerus," *Biology Letters,* vol. 8, pp. 609–612 (2012).

WHY DID SEXUAL REPRODUCTION EVOLVE, AND HOW DID THIS HAPPEN?

The Costs of Reproducing Sexually

Charles Darwin defines sexual selection in *On the Origin of Species,* John Murray (1859), pp. 87–90, and describes it in more detail in *The Descent of Man and Selection in Relation to Sex,* John Murray (1871), vol. 1, pp. 256–257. Jerry Coyne gives a more modern summary of the topic in *Why Evolution Is True,* Viking Penguin (2009). Michael Ryan shows that wild female túngara frogs choose mates by the nature of their calls in "Female mate choice in a Neotropical frog," *Science,* vol. 209, pp. 523–525 (1980). J. Barske et al. describe the acrobatics female manakins demand from their suitors in "Female choice for male motor skills," *Proceedings of the Royal Society,* series B, vol. 278, pp. 3523–3528 (2011). Jared Diamond describes the standards a male's bower must meet to attract females in "Experimental study of bower decoration by the bowerbird *Amblyornis inornatus,* using colored poker chips," *American Naturalist,* vol. 131, pp. 631–653 (1988). Even peahens that choose peacocks with the most elaborate tails are choosing good genes for their young: M. Petrie, "Improved growth and survival of offspring of peacocks with more elaborate trains," *Nature,* vol. 371, pp. 598–599 (1994). R. O. Prum insists, probably rightly, that in many birds, sexual selection is based on female choices driven by their aesthetics, their appreciation of beauty for its own sake: see pp. 17–225 in *The Evolution of Beauty,* Doubleday (2017). Michael Ryan discusses female aesthetics and from whence they evolve in *A Taste for the Beautiful: The Evolution of Attraction,* Princeton University Press (2018). William Eberhard surveys the variety of post-copulatory sexual selection in *Sexual Selection and Animal Genitalia,* Harvard University Press (1985) and *Female Control: Sexual Selection by Cryptic Female Choice,* Princeton University Press (1996). Every sexually produced organism grows from a single fertilized egg-cell: R. K. Grosberg and R. R. Strathmann, "One cell, two cell, red cell, blue cell: The persistence of a unicellular stage in multicellular life histories," *Trends in Ecology and Evolution,* vol. 13, pp. 112–116 (1998).

The Overriding Advantage of Sexual Reproduction

E. J. H. Corner explained the evolution of sexual reproduction, and its 50 percent cost, on pp. 20–22 and p. 86 of *The Life of Plants,* World Press (1964): the quotation is from p. 22 of this book. The advantage of sexual reproduction—recombining genes—is explained by J. Felsenstein, "The evolutionary advantage of recombination," *Genetics,* vol. 78, pp. 737–756 (1974) and J. Felsenstein and S. Yokoyama, "The evolutionary advantage of recombination II. Individual selection for recombination," *Genetics,* vol. 83, pp. 845–859 (1976). Felsenstein clearly understood that sexual reproduction and genetic recombination allowed new alleles to be tested more nearly by their own merits than by the accident of the genotype where they were born—an explanation experimentally confirmed by M. J. McDonald et al., "Sex speeds adaptation by altering the dynamics of molecular evolution," *Nature,* vol. 531, pp. 233–236 (2016). More recent surveys of explanations of why sexual reproduction evolved are given by J. A. G. M. de Visser and S. F. Elena, "The evolution of sex: Empirical insights into the roles of epistasis and drift," *Nature Reviews Genetics,* vol. 8, pp. 139–149 (2007). In *Homology, Genes, and Evolutionary Innovation,* Princeton University Press (2014), G. P. Wagner describes various forms of modularity in gene action that facilitate adaptive evolution. L. T. Morran et al. assessed the selective advantage of sex for the soil nematode *Caenorhabditis* in "Mutation load and rapid adaptation favor outcrossing over self-fertilization," *Nature,* vol. 462, pp. 350–352 (2009). The evolutionary relationship between transformation and meiosis is explained by H. and C. Bernstein, "Evolutionary origin of recombination during meiosis," *BioScience,* vol. 60, pp. 498–505 (2010).

WHAT MAKES NATURAL SELECTION ON SELFISH GENES FAVOR ADAPTED INDIVIDUALS?

Leigh outlines the advantage of fair meiosis in "The group selection controversy," *Journal of Evolutionary Biology,* vol. 23, pp. 6–19 (2010), p. 10. R. C. Lewontin and L. C. Dunn describe the array of **t**-alleles in house mice in "The evolutionary dynamics of a polymorphism in the house mouse," *Genetics,* vol. 45, pp. 705–722 (1960). J. F. Crow describes other "cheater" alleles in "Genes that violate Mendel's rules," *Scientific American,* vol. 240, no. 2, pp. 134–146 (1979). How selection at another chromosome can restore fair meiosis is analyzed by T. Prout et al., "Population genetics of modifiers of meiotic drive I. The solution of a special case and some general implications," *Theoretical Population Biology,* vol. 4, pp. 446–465 (1973). Adam Smith suggests that competition will be kept fair if disinterested bystanders combine to suppress unfair competitors in Part II, Section ii, chapter 2, ¶ 1 of *The Theory of Moral Sentiments,* A. Millar (1759). He argues that when each individual pursues his own self-interest by fair means, society benefits as if by an "invisible hand" in Book IV, chapter II of *An Inquiry into the Nature and Causes of the Wealth of Nations,* W. Strahan and T. Cadell (1776). Robert MacArthur shows how selection on the autosomes favors devoting equal effort to offspring of each sex, so that autosomal mutants would be favored that mitigate male bias in sex ratio in "Population effects of natural selection," *American Naturalist,* vol. 95, pp. 195–199 (1961), and A. B. Carvalho et al. demonstrate this experimentally in "An experimental demonstration of Fisher's principle: Evolution of sexual proportion by natural selection," *Genetics,* vol. 148, pp. 719–731 (1998). D. O. Conover and D. A. Van Voorhees, "Evolution of a balanced sex ratio by frequency-dependent selection in a fish," *Science,* vol. 250, pp. 1556–1558 (1990) demonstrate adaptive adjustment of temperature-dependent sex ratio in Atlantic silversides.

Eukaryotes evolved genomes organized for adaptive evolution two billion years ago, but animals triggered rapid adaptive evolution and diversification over a billion years later, well after there was enough oxygen to allow it: N. J. Butterfield, "Macroevolution and macroecology through deep time," *Palaeontology,* vol. 50, pp. 41–55 (2007) and "Oxygen, animals and bioturbation: An updated account," *Geobiology,* vol. 16, pp. 3–16 (2018).

Chapter 9. The Processes of Evolution

The initial quotation is from J. B. S. Haldane and J. S. Huxley, *Animal Biology,* Oxford University Press (1927), pp. 234–235.

DIVERGENCE AND THE ORIGIN OF SPECIES

Jerry Coyne discusses how one species becomes two in *Why Evolution Is True,* Viking Penguin (2009), pp. 168–189, as do James Sobel et al. in "The biology of speciation," *Evolution,* vol. 64, pp. 295–315 (2009). H. Allen Orr shows how substitution of incompatible alleles in two isolated halves of a population leads to speciation in "The population genetics of speciation: The evolution of hybrid incompatibilities," *Genetics,* vol. 139, pp. 1805–1813 (1995). Examples of "accidental" speciation are given in N. Knowlton et al., "Divergence in proteins, mitochondrial DNA, and reproductive compatibility across the Isthmus of Panama," *Science,* vol. 260, pp. 1629–1632 (1993) and H. A. Lessios, "Possible prezygotic reproductive isolation in sea urchins separated by the Isthmus of Panama," *Evolution,* vol. 38, pp. 1144–1148 (1984). On the other hand, R. W. Rubinoff and I. Rubinoff found that intertidal gobies from the two sides of the isthmus could interbreed, in "Geographic and reproductive isolation in Atlantic and Pacific populations of Panamanian *Bathygobius,*" *Evolution,* vol. 25, pp. 88–97 (1971), and H. A. Lessios and C. W. Cunningham found the same for a split pair of *Echinometra* sea urchins, "Gametic incompatibility between species of the sea urchin Echinometra on the two sides of the isthmus of Panama," *Evolution,* vol. 44, pp. 933–941 (1990). L. Cortés-Ortiz et al. discuss speciation of howler monkeys in MesoAmerica and South America in "Molecular systematics and biogeography of the Neotropical monkey genus *Alouatta,*" *Molecular Phylogenetics and Evolution,* vol. 26, pp. 64–81 (2003). R. A. Fisher outlines the role of trade-offs in speciation in *Genetical Theory of Natural Selection,* Oxford University Press (1930), pp. 125–128.

Speciation by Choosing Mates with the Same Way of Life

Chris Jiggins et al. showed how the two butterfly species *Heliconius melpomene* and *Heliconius cydno* diverged in "Reproductive isolation caused by color pattern mimicry," *Nature,* vol. 411, pp. 302–305 (2001). Douglas Schemske and H. D. Bradshaw, Jr., describe how one species of monkeyflower, *Mimulus,* became two in "Pollinator preference and the evolution of floral traits in monkeyflowers (*Mimulus*)," *Proceedings of the National Academy of Sciences, USA,* vol. 96, pp. 11910–11915 (1999). The trade-offs driving their divergence are outlined on pp. 1521 and 1528 of J. Ramsey et al., "Components of reproductive isolation between the monkeyflowers *Mimulus lewisii* and *M. cardinalis* (Phrymaceae)," *Evolution,* vol. 57, pp. 1520–1534 (2003).

Innovation, Hybrid Inviability, and Speciation

Christopher Wheat et al., "From DNA to fitness differences: Sequences and structures of adaptive variants of *Colias* phosphoglucose isomerase (PGI)," *Molecular Biology and Evolution,* vol. 23, pp. 499–512 (2006) and Christopher Wheat and Ward Watt, "A mitochondrial-DNA-based phylogeny for some evolutionary-genetic model species of *Colias* butterflies (Lepidoptera: Pieridae)," *Molecular Phylogenetics and Evolution,* vol. 47, pp. 893–902 (2008), describe the trade-off sulphur butterflies face between montane and lowland life and the divergence of lowland from montane *Colias.*

Speciation Driven by Behavioral Change

Richard Wrangham outlines the stages of evolution of human beings from the apelike *Australopithecus* in *Catching Fire: How Cooking Made Us Human,* Basic Books (2009). In "Impact of meat and Lower Palaeolitihic food processing techniques on chewing in humans," *Nature,* vol. 531, pp. 500–503 (2016), K. Zink and D. E. Lieberman observe that there is little evidence of cooking with fire before 500,000 years ago and argue that *Homo erectus* made their food more edible by pounding the underground tubers (like yams) they found, and slicing meat into small pieces with stone knives.

NATURAL SELECTION, AN INDIVIDUAL'S ADVANTAGE, AND ITS GROUP'S GOOD

Christian Ziegler and Egbert Leigh outline the problems and benefits of group life in *A Magic Web,* Oxford University Press (2002), pp. 225–230.

A Challenge of Group Life: How to Cope with Cheaters?

E. G. Leigh, Jr., outlines how cooperative societies control cheaters in "The group selection controversy," *Journal of Evolutionary Biology,* vol. 23, pp. 6–19 (2010).

Tragedies of the Commons

Garrett Hardin first described such tragedies in "The tragedy of the commons," *Science,* vol. 162, pp. 1243–1248 (1968). F. Berkes et al. outline ways to avert these tragedies, and their dangers, in "The benefits of the commons," *Nature,* vol. 340, pp. 91–93 (1989). Robert Netting shows how Swiss villagers manage commonlands in "What Alpine peasants have in common: Observations on communal tenure in a Swiss village," *Human Ecology,* vol. 4, pp. 135–146 (1976). Evils arising when villagers are not allowed to control access to their commonlands are described on pp. 739–743 of M. D. Subash Chandran and Madhav Gadgil, "State forestry and the decline in food resources in the tropical forests of Uttara Kannada, southern India," pp. 733–744 in C. M. Hladik et al., eds. *Tropical Forests, People and Food,* UNESCO (1993).

Cooperating with Relatives: Kin Selection

How kin selection works is described by A. F. G. Bourke, *Principles of Social Evolution,* Oxford University Press (2011), see especially pp. 28–73.

The natural history and solitary versus social behavior of nocturnal sweat bees, *Megalopta,* are discussed in A. R. Smith et al., "Survival and productivity benefits to social nesting in the sweat bee *Megalopta genalis* (Hymenoptera: Halictidae)," *Behavioral Ecology and Sociobiology,* vol. 61, pp. 1111–1120 (2007) and W. T. Wcislo and S. Tierney, "Behavioural environments and niche construction: The evolution of dim-light foraging in bees," *Biological Reviews,* vol. 84, pp. 19–37 (2009). Karen Kapheim et al. suggest that daughters must be "bullied" into working in "Kinship, parental manipulation and the evolutionary origin of eusociality," *Proceedings of the Royal Society,* series B, vol. 282, article # 20142886 (2015).

Social behavior in the wasp *Polistes canadensis* is discussed in Mary Jane West Eberhard, "Dominance relations in *Polistes canadensis* (L.), a tropical social wasp," *Monitore zoologico italiano,* new series, vol. 20, pp. 263–281 (1986); T. Giray et al., "Juvenile hormone, reproduction, and worker behavior in the neotropical social wasp *Polistes canadensis,*" *Proceedings of the National Academy of Sciences, USA,* vol. 102, pp. 3330–3335 (2005) and S. Sumner et al., "Radio-tagging technology reveals extreme nest-drifting behavior in a eusocial insect," *Current Biology,* vol. 17, pp. 140–145 (2007).

T. D. Seeley shows how kin selection shapes honeybee societies in *The Wisdom of the Hive,* Harvard University Press (1995), pp. 7–16. The contest between egg-laying workers and the workers that actually help their queen is analyzed by P. K. Visscher in "Reproductive conflict in honeybees: A stalemate of worker egg-laying and policing," *Behavioral Ecology and Sociobiology,* vol. 39, pp. 237–244 (1996). K. J. Loope et al. show that workers eat even full sisters' eggs in "No facultative worker policing in the honey bee (*Apis mellifera* L.)," *Naturwissenchaften,* vol. 100, pp. 473–477 (2013).

How and Why Do Unrelated Animals Sometimes Cooperate for Their Common Good?

The quotation about the Brigand Chih is from Arthur Waley, *Three Ways of Thought in Ancient China,* George Allen and Unwin (1939), p. 74.

Social nesting in greater anis and its causes is explained by Christina Riehl, then a graduate student at Princeton University doing dissertation research on them, in "Living with strangers: Direct benefits favor non-kin cooperation in a communally nesting bird," *Proceedings of the Royal Society,* series B, vol. 278, pp. 1728–1735 (2011) and "Egg ejection risk and hatching asynchrony predict egg mass in a communally breeding cuckoo, the Greater Ani (*Crotophaga major*)," *Behavioral Ecology,* vol. 21, pp. 676–683 (2010).

Egbert Leigh and Thelma Rowell suggest how the common interest of a group's members in their group's welfare promotes cooperation among members of a group of social birds or mammals in "The evolution of mutualism and other forms of harmony at various levels of biological organization," *Écologie,* vol. 26, pp. 131–158 (1995). A group of white-faced monkeys in its own territory enjoys a "home-court advantage" in a fight with intruders: M. Crofoot et al., "Interaction location outweighs the competitive advantage of numerical superiority in *Cebus capucinus* intergroup contests," *Proceedings of the National Academy of Sciences, USA,* vol. 105, pp. 577–581 (2008). Margaret Crofoot assesses the importance of winning such fights in "The cost of defeat: Capuchin groups travel further, faster and later after losing conflicts with neighbors," *American Journal of Physical Anthropology,* vol. 152, pp. 79–85 (2013).

Contrasting views of cooperation among a group's chimpanzees are provided by Frans de Waal, *Good Natured: The Origins of Right and Wrong in Humans and Other Animals,* Harvard University Press (1996), based on a large group of captive chimps, and M. N. Muller and J. C. Mitani, "Conflict and cooperation in wild chimpanzees," *Advances in the Study of Behavior,* vol. 35, pp. 275–331 (2005), which reviews studies of wild animals. C. M. Hladik's observations of wild chimpanzees in Gabon suggest that group members at least sometimes enforce their common interest in intra-group harmony.

COOPERATION AMONG MEMBERS OF DIFFERENT SPECIES: MUTUALISM

Factors favoring mutualism between members of different species are discussed by C. Ziegler and E. G. Leigh in *A Magic Web,* Oxford University Press (2002), pp. 62–84, and E. G. Leigh, "The evolution of mutualism," *Journal of Evolutionary Biology,* vol. 23, pp. 2507–2528 (2010). Mutualism is vital to natural ecosystems: pp. 211–212 of Leigh, *Tropical Forest Ecology.* Darwin remarked that a character evolved exclusively for the benefit of another species would be fatal to his theory and incompatible with natural selection in *The Origin of Species* (1859), p. 201. Martin Moynihan discusses the advantages of birds of many species joining in flocks in "The organization and probable evolution of some mixed species flocks of Neotropical birds," *Smithsonian Miscellaneous Collections,* vol. 143, pp. 1–140 (1962) and *Geographic Variation in Social Behavior and in Adaptations to Competition among Andean Birds,* Nuttall Ornithological Club (1979). Antwren flocks on Barro Colorado Island are described in Judy Gradwohl and Russell Greenberg, "The formation of antwren flocks on Barro Colorado Island, Panama," *Auk,* vol. 97, pp. 385–395 (1980) and R. Greenberg and J. Gradwohl, "A comparative study of the social organization of antwrens on Barro Colorado Island, Panama," *Ornithological Monographs,* vol. 36, pp. 845–855 (1985).

DEFENDING COOPERATORS AGAINST CHEATERS

Mutualists can cheat each other: E. A. Herre et al., "The evolution of mutualism: Exploring the path between conflict and cooperation," *Trends in Ecology and Evolution,* vol. 14, pp. 49–53 (1999). Cheating is suppressed in many ways: Leigh, "The evolution of mutualism," *Journal of Evolutionary Biology,* vol. 23, pp. 2507–2528 (2010).

Brief-Exchange Mutualisms

F. P. Schiestl and P. M. Schlüter show how orchids choose their pollinators and how their choice affects orchid diversity in "Floral isolation, specialized pollination, and pollinator behavior in orchids," *Annual Review of Entomology,* vol. 54, pp. 425–446 (2009). A pollinator-cheating orchid is described in Manfred Ayasse et al., "Evolution of reproductive strategies in the sexually deceptive orchid *Ophrys sphegodes:* How does flower-specific variation of odor signals influence reproduction?" *Evolution,* vol. 54, pp. 1995–2006 (2000). D. W. Roubik et al. describe how nectar robbers can reduce flowering plant reproduction in "Roles of nectar robbers in reproduction of the tropical treelet *Quassia amara* (Simaroubaceae)," *Oecologia* vol. 66, pp. 161–167 (1985).

Symbioses: Cooperators that Live Together

Leigh discusses what maintains the mutualism between leaf-cutter ants and its fungi on p. 2512 of "The evolution of mutualism," *Journal of Evolutionary Biology,* vol. 23, pp. 2507–2528 (2010). E. A. Herre evaluates factors affecting the impact of parasitic nematodes on the fig wasps they infest in "Population structure and the evolution of virulence in nematode parasites of fig wasps," *Science,* vol. 259, pp. 1442–1445 (1993). E. A. Herre et al. describe the symbiosis between figs and their pollinating wasps, and how it is enforced, in "Evolutionary ecology of figs and their associates: Recent progress and outstanding puzzles," *Annual Review of Ecology, Evolution and Systematics,* vol. 39, pp. 439–458 (2008). G. E. D. Oldroyd describes how plants summon rhizobial bacteria from the soil in "Speak, friend, and enter: Signaling systems that promote beneficial symbiotic associations in plants," *Nature Reviews Microbiology,* vol. 11, pp. 252–263 (2013). E. Toby Kiers et al. show how soybean plants ensure that the nitrogen-fixing bacteria they feed and house continue to supply them with ammonia in "Host sanctions and the legume-rhizobium mutualism," *Nature,* vol. 425, pp. 78–81 (2003). The symbiosis between reef corals and their zooxanthellae is outlined by K. Koike et al., "Octocoral chemical signaling selects and controls dinoflagellate symbionts," *Biological Bulletin,* vol. 207, pp. 80–86 (2004); A. Baker, "Flexibility and specificity in coral-algal symbiosis: Diversity, ecology and biogeography of *Symbiodinium,*" *Annual Review of Ecology, Evolution and Systematics,* vol. 34, pp. 661–689 (2003); and M. del C. Gómez-Cabrera et al., "Acquisition of symbiotic dinoflagellates (*Symbiodinium*) by juveniles of the coral *Acropora longicyathus,*" *Coral Reefs,* vol. 27, pp. 219–226 (2008).

D. H. Janzen describes mutualism between swollen-thorn acacias and the ants that protect them in "Coevolution of mutualism between ants and acacias in Central America," *Evolution,* vol. 20, pp. 249–275 (1966) and an ant that parasitizes these acacias in "*Pseudomyrmex nigropilosa:* A parasite of a mutualism," *Science,* vol. 188, pp. 936–937 (1975). Mutualisms among sponges and their subversion are described by J. L. Wulff in "Mutualisms among species of coral reef sponges," *Ecology,* vol. 78, pp. 146–159 (1997) and "Life-history differences among coral reef sponges promote mutualism or exploitation of mutualism by influencing partner fidelity feedback," *American Naturalist,* vol. 171, pp. 597–609 (2008). Leigh summarizes work on bobtail squid and their bacteria on pp. 2513–2514 of "The evolution of mutualism," *Journal of Evolutionary Biology,* vol. 23, pp. 2507–2528 (2010). For more detail, see M. J. McFall-Ngai, "Consequences of evolving with bacterial symbionts: Insights from the squid-vibrio associations," *Annual Review of Ecology and Systematics,* vol. 30, pp. 235–256 (1999), and M. J. McFall-Ngai and E. G. Ruby, "Sepiolids and vibrios: When first they meet," *BioScience,* vol. 48, pp. 257–265 (1998).

MAJOR TRANSITIONS OF EVOLUTION

The term "evolution the tinker" arose from a paper by F. Jacob, "Evolution and tinkering," *Science,* vol. 196, pp. 1161–1166 (1977). The quotation is from p. 235 of J. B. S. Haldane and J. S. Huxley, *Animal Biology,* Oxford University Press (1927).

A. F. G. Bourke discusses how selection achieved the major evolutionary transitions in *Principles of Social Evolution,* Oxford University Press (2011), pp. 4–21, and distinguishes egalitarian from fraternal transitions on p. 7. How archaeans transformed bacteria they engulfed into mitochondria, is detailed in N. Lane & W. Martin, "The energetics of genome complexity," *Nature,* vol. 467, pp. 929–934 (2010) and N. Lane, *The Vital Question: Energy, Evolution and the Origin of Complex Life,* Norton (2015). E. G. Leigh, Jr., describes the domestication of these symbionts in "When does the good of the group override the advantage of the individual," *Proceedings of the National Academy of Sciences, USA,* vol. 80, pp. 2985–2989 (1983) and on p. 259 of "Genes, bees and ecosystems: The evolution of a common interest among individuals," *Trends in Ecology and Evolution,* vol. 6, pp. 257–262 (1991). R. Grosberg and R. Strathmann discuss how multicellularity evolves in "The evolution of multicellularity: A minor major transition?" *Annual Review of Ecology, Evolution and Systematics,* vol. 38, pp. 621–654 (2007), and explain why slime mold aggregates, with cells of different genotypes, are so vulnerable to cheating on p. 627. A. H. Knoll discusses the origin of complex multicellular organisms in "The multiple origins of complex multicellularity," *Annual Review of Earth and Planetary Sciences,* vol. 39, pp. 217–239 (2011). M. W. Kirschner and J. C. Gerhart show how chemical gradients lay the foundations for developing different tissues and organs when all cells have the same genes in *The Plausibility of Life,* Yale University Press (2005), especially 209–215. See also S. B. Carroll, *Endless Forms Most Beautiful,* Norton (2005). Leo Buss shows how metazoans control rogue mutants in *The Evolution of Individuality,* Princeton University Press (1987). Genetic conflicts left over from a major transition of evolution where wholes are assembled from formerly independent parts show how natural selection effected that transition: pp. 26–29 of Leigh, "Levels of selection, potential conflicts, and their resolution: The role of the 'common good,'" in L. Keller, ed., *Levels of Selection in Evolution,* Princeton University Press (1999). J. Maynard Smith describes the two-legged goat in *The Theory of Evolution,* Penguin Books (1958), pp. 279–280. The final quotation is taken from Sewall Wright, "Evolution in Mendelian populations," *Genetics,* vol. 16, pp. 97–159. Of the founders of mathematical evolutionary theory, Ronald Fisher, J. B. S. Haldane and S. Wright, Wright was most concerned with how genes affect the development and function of their organisms.

Chapter 10. The Last Transition: How Thought and Language Evolved

This chapter's argument is based on Konrad Lorenz, *Behind the Mirror,* Methuen (1977). He discusses the impact of the origin of life on pp. 27–28, cultural evolution on pp. 172–173, and how language eases teaching young about threats on p. 160; the quotations are from p. 171 and p. 24.

CULTURAL VERSUS GENETIC INHERITANCE: PROS AND CONS

G. J. Vermeij and E. G. Leigh, Jr., discuss these pros and cons in "Natural and human economies compared," *Ecosphere,* vol. 2, no. 4, Article 39 (2011). Lorenz remarks on the greater adaptability but lower reliability of cultural evolution on p. 174, and the danger of cultural evolution outrunning human nature on p. 191, in *Behind the Mirror.* R. F. Ewer shows how meerkats teach their young how to handle difficult prey in "The 'instinct to teach,'" *Nature,* vol. 222, p. 698 (1969).

STEPS TOWARD CONCEPTUAL THOUGHT AND LANGUAGE: KNOWING OBJECTIVELY

The opening quotations are from p. 1 and p. 4 of Lorenz, *Behind the Mirror.* J.-P. Changeux and P. Ricoeur, *What Makes Us Think?* Princeton University Press (2000), is an extraordinary and informative dialogue between a neurobiologist and a moral philosopher on what neurobiology can tell us about how minds work. Lorenz discusses the many ingredients of conceptual thought on pp. 113–164 of *Behind the Mirror.* F. Jacob coined the term "evolution the tinker" in *Science,* vol. 196, pp. 1161–1166 (1977).

Why Do Animals Need Objective Knowledge, and How Do They Get It?

The quotations in the first paragraph are from F. Jacob, *The Possible and the Actual,* Pantheon Books (1982), p. 56 and p. 55. J. Monod discusses the function of brains and why they are organized to attain objective knowledge of their environment in *Chance and Necessity,* Vintage Books (1972), p. 154. Paul Ricoeur remarks on living beings organizing their world by "choosing meaningful signals" on p. 44 of J.-P. Changeux and P. Ricoeur, *What Makes Us Think?* Princeton University Press (2000). Monod explains the reaction of frogs to still vs. moving specks on pp. 150–151 of *Chance and Necessity:* his account is based on J. Y. Lettvin et al., "What the frog's eye tells the frog's brain," *Proceedings of the IRE,* vol. 47, pp. 1940–1951 (1959). A. Schmidt and H. Römer show how crickets distinguish the calls of males of their species from the many other night noises of a tropical forest in "Solution to the cocktail party problem in insects . . . ," *PLoS One,* vol. 6 (issue 12), e28593 (2011). W. E. Conner and A. J. Corcoran show how various insects detect (and cope with) bats seeking to eat them in "Sound strategies: The 65-million-year-old battle between bats and insects," *Annual Review of Entomology,* vol. 57, pp. 21–39 (2012). Lorenz mentions the ability we share with honeybees of distinguishing the "true" color of an object on pp. 116–117 of *Behind the Mirror.* Lorenz discusses fixed motor patterns on pp. 55–60, and the graylag goose example on p. 135, and the role of splitting fixed motor patterns into individually controlled segments in evolving objective thought on pp. 132–144 of the same book. J.-H. Fabre provides insect examples of fixed motor patterns in "Aberrations of instinct," in *The Mason Wasps,* Dodd, Mead (1919), pp. 106–132. Lorenz discusses the unconscious abstractions involved in perception and their implications on pp. 114–120 of *Behind the Mirror.* H. Weyl discusses how such abstracting perception resembles that of physicists on p. 199, H. Weyl, *Mind and Nature,* Princeton University Press (2009). D. L. Cheney and R. M. Seyfarth remark that vervet monkeys distinguish different predators by different alarm calls on pp. 102–113 of *How Monkeys See the World,* University of Chicago Press (1990). F. Jacob describes the crucial importance of the ability to resolve what one sees into discrete objects on pp. 57–58 of *The Possible and the Actual,* Pantheon Books (1982). Lorenz discusses how animals "situate" objects in space, and implications for how we speak and think, in *Behind the Mirror,* pp. 120–129. F. R. Wilson outlines the role of hands in *The Hand: How Its Use Shapes the Brain, Language, and Human Culture,* Vintage Books (1999).

How Social Life Can Help Thought and Language Evolve

Alison Jolly discusses why social animals need social knowledge in "Lemur social behavior and primate intelligence," *Science,* vol. 153, pp. 501-506 (1966). D. L. Cheney and R. M. Seyfarth discuss baboons' mental powers in *Baboon Metaphysics,* University of Chicago Press (2007), pp. 90–283. Ray Jackendoff discusses how quickly children learn to speak on pp. 102–104, the similar principles of spoken and sign language on p. 98, and how much sign language apes can learn, and why they cannot learn more, on pp. 135–139 of *Patterns in the Mind,* Basic Books

(1994). F. R. Wilson remarks that the same lesions block speech and sign language on p. 200 of *The Hand,* Vintage Books (1999). Colin McGinn argues that in human beings, sign preceded spoken language on pp. 39–59 of *Prehension: The Hand and the Emergence of Humanity,* MIT Press (2015). Ray Jackendoff shows how social life shapes thought and language in *Patterns in the Mind,* pp. 204–222. Darwin showed that human facial expressions normally reveal the same emotion the world over in *The Expression of the Emotions in Man and Animals,* John Murray (1872).

The Mystery of Consciousness

The opening quote is from p. 283 of H. Weyl, *The Philosophy of Mathematics and Natural Science,* Princeton University Press (1949). His remark implies that consciousness underlies "theory of mind." Monod remarked on consciousness as simulator in *Chance and Necessity,* p. 158. Lorenz discusses the box-moving orangutan in *Behind the Mirror,* p. 128. Stanislas Dehaene shows how one can study consciousness experimentally in *Consciousness and the Brain,* Penguin Books (2014).

How Play Can Bring Forth Objective Knowledge

Lorenz discusses the significance of curiosity, play, and exploratory behavior on pp. 144–151, and the significance of apes watching their hands when moving or manipulating, on pp. 150–151, of *Behind the Mirror:* the quotation relating exploratory play to scientific research is from p. 149. J. P. Changeux suggests a neurological basis for the drive to discover on pp. 68 and 73–74 of *L'Homme de verité,* Odile Jacob (2002). A fine Dutch historian, Johan Huizinga, discusses the role of play in animal life and human culture in *Homo Ludens: A Study of the Play Element of Culture,* Roy (1950): the quotation about playing dogs is from p. 1. Mathematics is even more like play than basic scientific research, a striving, as in poetry, to reveal beautiful patterns: G. H. Hardy, *A Mathematician's Apology,* 2nd edition, Cambridge University Press (1967), pp. 85–86. S. Dehaene discusses the number sense of other vertebrates in *La Bosse des maths,* 2nd ed., Odile Jacob (2010). Exploratory play discovers truths like the existence of only 17 types of wallpaper symmetry, that are independent of our culture or even our existence, as H. Weyl showed in *Symmetry,* Princeton University Press (1951); and A. Connes shows in J.-P. Changeux and A. Connes, *Conversations on Mind, Matter and Mathematics,* Princeton University Press (1995), p. 5. J. A. Wheeler discusses how stars and planets curve space, and how curved space affects their motion, in *A Journey into Gravity and Spacetime,* Scientific American Library (1990). Eugene Wigner discusses mathematics as a language of physics and a guide to its study in "The unreasonable effectiveness of mathematics in the natural sciences," pp. 222–237 in Eugene Wigner, *Symmetries and Reflections: Scientific Essays,* Indiana University Press (1967). The last quotation is from p. 252 of G. E. Hutchinson, *The Kindly Fruits of the Earth: Reflections of an Embryo Ecologist,* Yale University Press (1979).

Chapter 11. What Have We Learned, and What Is Still Unknown?

FRONTIERS OF GROSS IGNORANCE

J. Lovelock, *The Ages of Gaia,* Oxford University Press (1988) and G. J. Vermeij, "A historical conspiracy: Competition, opportunity, and the emergence of direction in history," *Cliodynamics,* vol. 2, pp. 187–207 (2011) discuss why natural communities are so productive and diverse. The first quotation is from D. W. Thompson. *On Growth and Form,* 2nd ed., Cambridge University Press (1942), pp. 14–15. The quote from Hermann Weyl is from *The Philosophy of Mathematics and Natural Science,* Princeton University Press (1949), p. 283. Stanislas Dehaene discusses the tendency of some psychologists and animal behaviorists to avoid consciousness on pp. 7–8 of *Consciousness and the Brain,* Penguin (2014), most of which is devoted to showing how focus on conscious access has amplified our understanding of consciousness. Jacques Monod describes what he considered the deep divide between objective knowledge and introspective consciousness in *Chance and Necessity,* Vintage Books (1972), p. 159.

Index

Page numbers in italics refer to illustrations